Lecture Notes in Mathematics

523

Sergio A. Albeverio · Raphael J. Høegh-Krohn
Sonia Mazzucchi

Mathematical Theory
of Feynman Path Integrals

An Introduction

2nd corrected and enlarged edition

 Springer

Authors

Sergio A. Albeverio
Department of Mathematics
University of Bonn
Wegelerstr. 6
53115 Bonn
Germany
albeverio@uni-bonn.de

Raphael J. Høegh-Krohn
University of Oslo
Blindern
Norway

Sonia Mazzucchi
Department of Mathematics
University of Trento
via Sommarive 14
38050 Povo (TN)
Italy
mazzucch@science.unitn.it

ISBN: 978-3-540-76954-5   e-ISBN: 978-3-540-76956-9
DOI: 10.1007/978-3-540-76956-9

Lecture Notes in Mathematics ISSN print edition: 0075-8434
ISSN electronic edition: 1617-9692

Library of Congress Control Number: 2008921370

Mathematics Subject Classification (2000): 81Q30, 81S40, 28C20, 46Nxx, 46T12, 47D08, 47G30, 60H05, 58D30, 35S30, 41A60, 34E05, 35Q40, 42A38, 81T45

Cover design: WMXDesign GmbH

Printed on acid-free paper

9 8 7 6 5 4 3 2 1

springer.com

Preface to the Second Edition

This second edition, unfortunately, had to be done without the contribution of Raphael Høegh-Krohn, who died on 28 January 1988. The authors of the present edition hope very much that the result of their efforts would had been appreciated by him. His beloved memory has been a steady inspiration to us. Since the appearance of the first edition many new developments have taken place. The present edition tries to take this into account in several ways, keeping however the basic structure and contents of the first edition. At that time the book was the first rigorous one to appear in the area and was written in a sort of pioneering spirit. In our opinion it is still valid as an introduction to all the work which followed; therefore in this second edition we preserve its form entirely (except for correcting some misprints and slightly improving some formulations). A chapter has been however added, in which many new developments are included. These concern both new mathematical developments in the definition and properties of the integrals, and new exciting applications to areas like low dimensional topology and quantized gauge fields. In addition we have added historical notes to each of the chapters and corrected several misprints of the previous edition. As for references, we have kept all those of the first edition, numbered from 1 to 56 (with the corresponding updating), and added new references (in alphabetic order).

We are very grateful to many coworkers, friends and colleagues, who inspired us in a number of ways. Special thanks are due to Philippe Blanchard, Zdzisław Brzeźniak, Luca Di Persio, Jorge Rezende, Jörg Schäfer, Ambar Sengupta, Ludwig Streit, Aubrey Truman, Luciano Tubaro, and Jean-Claude Zambrini. We also like to remember with gratitude the late Yuri L. Daleckii and Michel Sirugue who gave important contributions to this area of research.

Trento,
June 2005

Sergio A. Albeverio
Sonia Mazzucchi

Preface to the First Edition

In this work we develop a general theory of oscillatory integrals on real Hilbert spaces and apply it to the mathematical foundation of the so-called Feynman path integrals of non-relativistic quantum mechanics, quantum statistical mechanics and quantum field theory. The translation invariant integrals we define provide a natural extension of the theory of finite dimensional oscillatory integrals, which has recently undergone an impressive development, and appear to be a suitable tool in infinite dimensional analysis. For example, on the basis of the present work, we have extended the methods of stationary phase, Lagrange immersions and corresponding asymptotic expansions to the infinite dimensional case, covering in particular the expansions around the classical limit of quantum mechanics. A particular case of the oscillatory integrals studied in the present work are the Feynman path integrals used extensively in physics literature, starting with the basic work on quantum dynamics by Dirac and Feynman, in the 1940s.

In the introduction, we give a brief historical sketch and some references concerning previous work on the problem of the mathematical justification of Feynman's heuristic formulation of the integral. However, our aim with the present publication was not to write a review work, but rather to develop from scratch a self-contained theory of oscillatory integrals in infinite dimensional spaces, in view of the mathematical and physical applications mentioned above.

The structure of the work is briefly as follows. It consists of nine chapters. Chapter 1 is the introduction. Chapters 2 and 4 give the definitions and basic properties of the oscillatory integrals, which we call Fresnel integrals or normalized integrals, for the cases where the phase function is a bounded perturbation of a non-degenerate quadratic form (positive in Chap. 2). Chapters 3 and 5–9 give applications to quantum mechanics, namely N-particle systems with bounded potentials (Chap. 3) and systems of harmonic oscillators with finitely or infinitely many degrees of freedom (Chaps. 5–9), with relativistic quantum fields as a particular case (Chap. 9).

This work appeared first as a Preprint of the Mathematics Institute of Oslo University, in October 1974.

The first named author would like to express his warm thanks to the Institute of Mathematics, Oslo University, for the friendly hospitality. He also gratefully acknowledges the financial support of the Norwegian Research Council for Science and the Humanities. Both authors thank Mrs. S. Cordtsen, Mrs. R. Møller and Mrs. W. Kirkaloff heartily for their patience and skill in typing the manuscript.

Oslo,   *Sergio A. Albeverio*
March 1976   *Raphael J. Høegh-Krohn*

Contents

Preface to the Second Edition V

Preface to the First Edition VII

1   Introduction 1

2   The Fresnel Integral of Functions on a Separable Real
    Hilbert Space... 9

3   The Feynman Path Integral in Potential Scattering 19

4   The Fresnel Integral Relative to a Non-singular Quadratic
    Form 37

5   Feynman Path Integrals for the Anharmonic Oscillator 51

6   Expectations with Respect to the Ground State
    of the Harmonic Oscillator............................... 63

7   Expectations with Respect to the Gibbs State
    of the Harmonic Oscillator............................... 69

8   The Invariant Quasi-free States 73

9   The Feynman History Integral
    for the Relativistic Quantum Boson Field................. 85

10 Some Recent Developments............................... 93
    10.1 The Infinite Dimensional Oscillatory Integral................ 93
    10.2 Feynman Path Integrals for Polynomially
         Growing Potentials 101
    10.3 The Stationary Phase Method and the Semiclassical
         Expansion 108

10.4 Alternative Approaches to Rigorous Feynman
Path Integrals.. 115
10.4.1 Analytic Continuation 115
10.4.2 White Noise Calculus Approach 116
10.4.3 The Sequential Approach........................... 120
10.4.4 The Approach via Poisson Processes 123
10.5 Recent Applications 124
10.5.1 The Schrödinger Equation with Magnetic Fields....... 124
10.5.2 The Schrödinger Equation with Time Dependent
Potentials 125
10.5.3 Phase Space Feynman Path Integrals 130
10.5.4 The Stochastic Schrödinger Equation 133
10.5.5 The Chern–Simons Functional Integral 136

References of the First Edition 141

References Added for the Second Edition 149

Analytic Index 173

List of Notations 177

1

Introduction

Feynman path integrals have been introduced by Feynman in his formulation of quantum mechanics [1].[1] Since their inception they have occupied a somewhat ambiguous position in theoretical physics. On one hand they have been widely and profitably used in quantum mechanics, statistical mechanics and quantum field theory, because of their strong intuitive, heuristic and formal appeal. On the other hand most of their uses have not been supported by an adequate mathematical justification. Especially in view of the potentialities of Feynman's approach as an alternative formulation of quantum dynamics, the need for a mathematical foundation has been broadly felt and the mathematical study of Feynman path integrals repeatedly strongly advocated, see e.g. [4]. This is, roughly speaking, a study of oscillating integrals in infinitely many dimensions, hence closely connected with the development of the theory of integration in function spaces, see e.g. [5]. The present work intends to give a mathematical theory of Feynman path integrals and to yield applications to non relativistic quantum mechanics, statistical mechanics and quantum field theory. In order to establish connections with previous work, we shall give in this introduction a short historical sketch of the mathematical foundations of Feynman path integrals. For more details we refer to the references, in particular to the review papers [6].

Let us first briefly sketch the heuristic idea of Feynman path integrals, considering the simple case of a non relativistic particle of mass m, moving in Euclidean space $\mathbb{R}^n$ under the influence of a conservative force given by the potential $V(x)$, which we assume, for simplicity, to be a bounded continuous real valued function on $\mathbb{R}^n$.

The classical Lagrangian, from which the classical Euler–Lagrange equations of motion follow, is

$$L\left(x, \frac{\mathrm{d}x}{\mathrm{d}t}\right) = \frac{m}{2}\left(\frac{\mathrm{d}x}{\mathrm{d}t}\right)^2 - V(x). \tag{1.1}$$

---

[1] A vivid account of the origins of the idea, influenced particularly by remarks of Dirac [2], has been given by Feynman himself in [3].

Hamilton's principle of least action states that the trajectory actually followed by the particle going from the point y, at time zero, to the point x at time t, is the one which makes the classical action, i.e. Hamilton's principal function,

$$S_t(\gamma) = \int_0^t L\left(\gamma(\tau), \frac{\gamma(\tau)}{d\tau}\right) d\tau \qquad (1.2)$$

stationary, under variations of the path $\gamma = \{\gamma(\tau)\}$, $0 \leq \tau \leq t$, with $\gamma(0) = y$ and $\gamma(t) = x$, which leave fixed the initial and end points y and x, and the time.

In quantum mechanics the state of the particle at time t is described by a function $\psi(x, t)$ which, for every t, belongs to $L_2(\mathbb{R}^n)$ and satisfies Schrödinger's equation of motion

$$i\hbar \frac{\partial}{\partial t}\psi(x, t) = -\frac{\hbar^2}{2m}\triangle\psi(x, t) + V(x)\psi(x, t), \qquad (1.3)$$

with prescribed Cauchy data at time $t = 0$,

$$\psi(x, 0) = \varphi(x), \qquad (1.4)$$

where $\triangle$ is the Laplacian on $\mathbb{R}^n$ and $\hbar$ is Planck's constant divided by 2π. The operator

$$H = -\frac{\hbar^2}{2m}\triangle + V(x), \qquad (1.5)$$

the Hamiltonian of the quantum mechanical particle, is self-adjoint on the natural domain of $\triangle$ and therefore $e^{-\frac{i}{\hbar}tH}$ is a strongly continuous unitary group on $L_2(\mathbb{R}^n)$. The solution of the initial value problem (1.3), (1.4) is

$$\psi(x, t) = e^{-\frac{i}{\hbar}tH}\varphi(x). \qquad (1.6)$$

From the Lie–Kato–Trotter product formula we have

$$e^{-\frac{i}{\hbar}tH} = s - \lim_{k \to \infty}\left(e^{-\frac{i}{\hbar}\frac{t}{k}V}e^{-\frac{i}{\hbar}\frac{t}{k}H_0}\right)^k, \qquad (1.7)$$

where

$$H_0 = -\frac{\hbar^2}{2m}\triangle. \qquad (1.8)$$

Assuming now for simplicity that φ is taken in Schwartz space $\mathcal{S}(\mathbb{R}^n)$, we have, on the other hand

$$e^{-\frac{i}{\hbar}tH_0}\varphi(x) = \left(2\pi i\frac{\hbar}{m}t\right)^{-\frac{n}{2}}\int e^{im\frac{(x-y)^2}{2\hbar t}}\varphi(y)dy, \qquad (1.9)$$

hence, combining (1.7) and (1.9)

$$e^{-\frac{i}{\hbar}tH}\varphi(x) = s - \lim_{k\to\infty} \left(2\pi i \frac{\hbar}{m}\frac{t}{k}\right)^{-\frac{kn}{2}} \int_{\mathbb{R}^{nk}} e^{-\frac{i}{\hbar}S_t(x_k,\dots,x_0)}\varphi(x_0)dx_0\dots dx_{k-1}$$

$$(1.10)$$

where by definition $x_k = x$ and

$$S_t(x_k,\dots,x_0) = \sum_{j=1}^{k}\left[\frac{m}{2}\frac{(x_j - x_{j-1})^2}{\left(\frac{t}{k}\right)^2} - V(x_j)\right]\frac{t}{k}. \tag{1.11}$$

The expression (1.10) gives the solution of Schrödinger's equation as a limit of integrals.

Feynman's idea can now be formulated as the attempt to rewrite (1.10) in such a way that it appears, formally at least, as an integral over a space of continuous functions, called paths. Let namely $\gamma(\tau)$ be a real absolutely continuous function on the interval $[0,t]$, such that $\gamma(\tau_j) = x_j$, $j = 0,\dots,k$, where $\tau_j = \frac{jt}{k}$ and $x_0,\dots,x_k$ are given points in $\mathbb{R}^n$, with $x_k = x$. Feynman looks upon $S_t(x_k,\dots,x_0)$ as a Riemann approximation for the classical action $S_t(\gamma)$ along the path γ,

$$S_t(\gamma) = \int_0^t \frac{m}{2}\left(\frac{d\gamma}{d\tau}\right)^2 d\tau - \int_0^t V(\gamma(\tau))\,d\tau. \tag{1.12}$$

Moreover when $k \to \infty$ the measure in (1.10) becomes formally $d\gamma = N\prod_{0\leq\tau\leq t}d\gamma(\tau)$, N being a normalization, so that (1.10) becomes the heuristic expression

$$\int_{\gamma(\tau)=x} e^{\frac{i}{\hbar}S_t(\gamma)}\varphi(\gamma(0))\,d\gamma, \tag{1.13}$$

where the integration should be over a suitable set of paths ending at time t at the point x. This is Feynman's path integral expression for the solution of Schrödinger's equation and we shall now review some of the work that has been done on its mathematical foundation.[2] Integration theory in spaces of continuous functions was actually available well before the advent of Feynman path integrals, particularly originated by Wiener's work (1923) on the Brownian motion, see e.g. [8]. It was however under the influence of Feynman's work that Kac [9] proved that the solution of the heat equation

$$\frac{\partial}{\partial t}f(x,t) = \sigma\triangle f(x,t) - V(x)f(x,t), \tag{1.14}$$

which is the analogue of Schrödinger's equation when t is replaced by $-it$, σ being diffusion's constant, can be expressed by

$$f(x,t) = \int e^{-\int_0^t V(\gamma(\tau)+x)d\tau}\varphi(\gamma(0)+x)\,dW(\gamma), \tag{1.15}$$

---

[2] For the physical foundation see the original work of Feynman and the book by Feynman and Hibbs [1]. Also e.g. [7].

where $dW(\gamma)$ is Wiener's measure for the Wiener, i.e. Brownian motion, process with variance $\sigma^2 d\tau$, defined on continuous paths[3] $\gamma(\tau)$, $0 \leq \tau \leq t$, with $\gamma(\tau) = 0$. Hence (1.15) is an expectation with respect to the normal unit distribution indexed by the real Hilbert space of absolutely continuous functions $\gamma(\tau)$, with norm $\|\gamma\|^2 = \int_0^t \left(\frac{d\gamma}{d\tau}\right)^2 d\tau$. From this we see that (1.15) can be formally rewritten as (1.13), with $\frac{1}{\hbar} S_t(\gamma)$ replaced by $-\frac{1}{2} \int_0^t \frac{1}{2\sigma} \left(\frac{d\gamma}{d\tau}\right)^2 d\tau - \int_0^t V(\gamma(\tau)) d\tau$. Thus (1.15) is a rigorous path integral (Wiener path integral) which plays for the heat equation a similar role as the Feynman path integral for the Schrödinger equation. This fact has been used [10] to provide a "definition by analytic continuation" of the Feynman path integral, in the sense that Feynman's path integral is then understood as the analytic continuation to purely imaginary t of the Wiener integral (1.15). The analogous continuation of the Wiener integral solution of the equation (1.14), with V replaced by iV, which corresponds to Schrödinger's equation with purely imaginary mass m, has been studied by Nelson [10] and allows to cover the case of some singular potentials. These definitions by analytic continuation, as well as the definition by the "sequential limit" (1.10),[4] have the disadvantage of being indirect in as much as they do not exhibit Feynman's solution (1.13) as an integral of the exponential of the action over a space of paths in physical space–time. In particular they are unsuitable for the mathematical realization of the original Dirac's and Feynman's ideas (see e.g. [1, 2])[5] about the approach to the classical limit $\hbar \to 0$, perhaps one of the most beautiful features of the Feynman path integral formalism. Namely (1.13) suggests that a suitable definition of the oscillatory integral should allow for the application of an infinite dimensional version of the method of stationary phase, to obtain, for $\hbar \to 0$, an asymptotic expansion in powers of $\hbar$, with leading term given by the path which makes $S_t(\gamma)$ stationary i.e., according to Hamilton's principle, the trajectory of classical motion. The definition of Feynman path integrals and more general oscillatory integrals in infinitely many dimensions which we give in this work is precisely well suited for this discussion, as shown in [41].[6]

Before we come however to our definition, let us make few remarks on other previous discussions of the mathematical foundations of Feynman path integrals. The attempt to define Feynman integral as a Wiener integral with purely imaginary variance meets the difficulty that the ensuing complex measure has infinite total variation (as first pointed out by Cameron [10], 1) and Daletskii [10], 2), in relation to a remark in [5]) and is thus unsuitable to define integrals like (1.13). For further remarks on this complex measure see [12].

A definition of Feynman path integrals for non relativistic quantum mechanics, not involving analytic continuation as the ones [10] mentioned before,

---

[3] Actually, Hölder continuous of index less than $1/2$, see e.g. [8].
[4] For the definition by a "sequential limit", in more general situations, see e.g. [11].
[5] See also e.g. the references given in [41] and [42].
[6] The results are also briefly announced in [42].

has been given by Ito [13]. We shall describe this definition in Chap. 2. Ito treated potentials $V(x)$ which are either Fourier transforms of bounded complex measures or of the form $c_\alpha x^\alpha$, with $\alpha = 1, 2$, $c_2 > O$. Ito's definition has been further discussed by Tarski [14]. Recently Morette-De Witt [15] has made a proposal for a definition of Feynman path integral, which has some relations with Ito's definition, but is more distributional rather than Hilbert space theoretical in character. The proposal suggests writing the Fourier transform of (1.13) as the "pseudomeasure"[7] $e^{-\frac{i}{\hbar}W}$, looked upon as a distribution acting on the Fourier transform of $e^{-\frac{i}{\hbar} \int_0^t V(\gamma(\tau))d\tau}$, provided this exists, where W is the Fourier transform of Wiener's measure with purely imaginary variance. This proposal left open the classes of functions V for which it actually works. Such classes follow however from Chap. 2 of the present work. Despite its, so far, incompleteness as to the class of allowed potentials, let us also mention a general attempt by Garczynski [16] to define Feynman path integrals as averages with respect to certain quantum mechanical Brownian motion processes, which generalize the classical ones. This approach has, incidentally, connections with stochastic mechanics [17], which itself would be worthwhile investigating in relation to the Feynman path formulation of quantum mechanics.[8]

Let us now make a corresponding brief historical sketch about the problem of the mathematical definition of Feynman path integrals in quantum field theory. They were introduced as heuristic tools by Feynman in [1] and applied by him to the derivation of the perturbation expansion in quantum electrodynamics. They have been used widely since then in the physical literature,

---

[7] A theory of related pseudomeasures has in-between been developed by Krée. See e.g. [43] and references therein.

[8] Besides the topics touched in this brief historical sketch of the mathematical study of Feynman path integrals of non relativistic quantum mechanics there are others we did not mention, either because they concern problems other than those tackled later in this work or because no clear cut mathematical results are available. Let us mention however three more areas in which Feynman path integrals have been discussed and used, at least heuristically.

(a) Questions of the relation between Feynman's quantization and the usual one: see e.g. [10],6), [31, 35].

(b) Feynman's path integrals on functions defined on manifolds other than Euclidean space, in particular for spin particles. Attempts using the sequential limit and analytic continuation approaches have been discussed to some extent, see e.g. [6],7), [36] and references given therein. For the analytic continuation approach there is available the well developed theory of Wiener integrals on Riemannian manifolds, see e.g. [37].

(c) As mentioned before, an important application of Feynman path integrals is in the discussion of the classical limit, where $\hbar \to 0$. In [41] we tackle this problem and we refer to this paper and [42] also for references (besides e.g. [1, 2, 5, 7, 38]).

see e.g. [18], also under the name of Feynman history integrals. We shall now
shortly give their formal expression. For more details see, besides the original
papers [1], also e.g. [18]. The classical formal action for the relativistic scalar
boson field is $S(\varphi) = S_0(\varphi) + \int_{\mathbb{R}^{n+1}} V(\varphi(\vec{x}, t)) \, d\vec{x} dt$, with

$$S_0(\varphi) = \frac{1}{2} \int\limits_{\mathbb{R}^{n+1}} \left[\left(\frac{d\varphi}{d\tau} \right)^2 - \sum_{i=1}^{n} \left(\frac{\partial\varphi}{\partial x_i} \right)^2 - m^2 \varphi^2 \right] d\vec{x} dt$$

where φ is a function of $\vec{x}, t$, m is a non negative constant, the mass of the field,
and V is the interaction. Similarly as in the case of a particle, the classical
solutions of the equations of motion is given by Hamilton's principle of least
action. The corresponding quantized system is formally characterized by the so
called time ordered vacuum expectation values $G(\vec{x}_1, t_1, \ldots, \vec{x}_k, t_k)$, formally
given, for $t_1 \leq \ldots \leq t_k$, $k = 1, 2, \ldots$, by the expectations of the products
$\Phi(\vec{x}_1, t_1) \ldots \Phi(\vec{x}_k, t_k)$ in the vacuum state, where $\Phi(\vec{x}, t)$ is the quantum field
(see e.g. [18], 7)). An heuristic expression for these quantities in terms of
Feynman history integrals is

$$G(\vec{x}_1, t_1, \ldots, \vec{x}_k, t_k) = T \left(\int e^{iS(\varphi)} \varphi(\vec{x}_1, t_1) \ldots \varphi(\vec{x}_k, t_k) \, d\varphi \right),$$

where T is the so called time ordering operator and the integrals are thought of
as integrals over a suitable subset of real functions φ on $\mathbb{R}^{n+1}$, see e.g. [18], 7).
A mathematical justification of this formula, or a related one, would actually
provide a solution of the well known problem of the construction of relativistic
quantum field theory. Somewhat in connection, in one way or the other, with
this problem, a large body of theory on integration in function spaces has been
developed since the fifties and we mention in particular the work by Friedrichs
[19], Gelfand [20], Gross [21] and Segal [22] and their associates, see also e.g.
[23]. With respect to the specific application to quantum field theory, more recently a study of models has been undertaken, see e.g. [24], in which either the
relativistic interaction is replaced by an approximate one, with the ultimate
goal of removing at a later stage the approximation, or physical space–time
is replaced by a lower dimensional one. We find here methods which parallel
in a sense those discussed above in relation with Schrödinger's equation and,
in a similar way as in that case, we can put these methods in connection with
the problem of giving meaning to Feynman path integral, although in this
case the connection is even a more indirect one as it was in the non relativistic case. We mention however these methods for their intrinsic interest.
The sequential approach based on Lie–Kato–Trotter formula has been used
especially in two space–time dimensional models particularly by Glimm, Jaffe
and Segal [25]. The analytic continuation approach, in which time is replaced
by imaginary time, is at the basis of the so called Euclidean–Markov quantum
field theory, pursued vigorously by Symanzik [26] and Nelson [27] and applied
particularly successfully, mostly in connection with the fundamental work of

Glimm and Jaffe, for local relativistic models in two space–time dimensions, with polynomial [24] or exponential interactions [28],[9] and in three space–time dimensions with space cut-off [29],[10] respectively in higher dimensions with ultraviolet cut-off interactions [30].[11] Much in the same way as for the heat, Schrödinger and stochastic mechanics equations, there are connections also with stochastic field theory [17], 4) - 7).[12]

Coming now to the Feynman history integrals themselves, it does not seem, to our knowledge, that any work has been done previous to our present work, as to their direct mathematical definition as integrals on a space of paths in physical space–time, except for the free case [14].

We shall now summarize briefly the content of the various sections of our work.

In Chap. 2 we introduce the basic definition for oscillating integrals on a separable real Hilbert space, which we call Fresnel integrals, and we establish their properties. In Chap. 3 this theory is applied to the definition of Feynman path integrals in non relativistic quantum mechanics. We prove that the heuristic Feynman path integral formula (1.13) for the solution of Schrödinger's equation can be interpreted rigorously as a Fresnel integral over a Hilbert space of continuous paths. In addition we derive corresponding formulae also for the wave operators and for the scattering operator.[13] In Chap. 4 we extend, in view of further applications, the definition of Fresnel integrals and give the properties of the new integral, called Fresnel integral relative to a given quadratic form. This theory is applied in Chap. 5 to the definition of Feynman path integrals for the n-dimensional anharmonic oscillator and in Chaps. 6 and 7 to the expression of expectations of functions of dynamical quantities of this anharmonic oscillator with respect to the ground state, respectively the Gibbs states [33] and quasifree states [34] of the correspondent harmonic oscillator.[14] In Chap. 8 we express the time invariant quasifree states on the Weyl algebra of an infinite dimensional harmonic oscillator by

---

[9] The Wightman axioms for a local relativistic quantum field theory (see e.g. [40]) have been proved, in particular.

[10] The space cut-off has now been removed [44].

[11] See also [45].

[12] We did not mention here other topics which have some relations to Feynman's approach to the quantization of fields, for much the same reason as in the preceding Footnote 8). For discussion of problems in defining Feynman path integrals for spinor fields see e.g. [36] and references given therein. For the problem of the formulation of Feynman path integral in general relativity see e.g. [39, 4],2), and references given there.

[13] Similar results hold for a system of N non relativistic quantum mechanical particles, moving each in d-dimensional space, interacting through a superposition of ν-body potentials ($\nu = 1, 2, \ldots$) allowed in particular to be translation invariant.

[14] The same results hold for a system of N anharmonic oscillators, with anharmonicities given by superpositions of ν-body potentials, as in the preceding footnote.

Feynman path integrals defined as Fresnel integrals in the sense of Chap. 4, and this also provides a characterization of such states.

Finally, in Chap. 9 we apply the results of Chap. 8 to the study of relativistic quantum field theory. For the ultra-violet cut-off models mentioned above [30] we express certain expectation values, connected with the time ordered vacuum expectation values, in terms of Feynman history integrals, again defined as Fresnel integrals relative to a quadratic form. We also derive the correspondent expressions for the expectations with respect to any invariant quasi-free state, in particular for the Gibbs states of statistical mechanics for quantum fields ([33]3)).

Notes

The introduction appears here unchanged from the one of the first edition which obviously took only into account developments up to the year of appearance (1975). Simultaneously to the appearance of the first edition of this book, a method of stationary phase for Feynman path integrals was developed [87] and Maslov's approach to Feynman path integrals via Poisson processes became known [38]. These and subsequent developments are discussed in Chap. 10. Concerning footnote 2 we might add the following more recent references (articles resp. books on Feynman path integrals and their applications, of general interest, not necessarily concerned with the rigorous approach discussed in the present book): [60, 75, 74, 111, 125, 161, 179, 192, 113, 208, 209, 210, 211, 228, 229, 251, 253, 264, 287, 315, 318, 323, 325, 354, 376, 378, 401, 404, 409, 458, 467, 250, 90].

2

The Fresnel Integral of Functions on a Separable Real Hilbert Space

We consider first the case of the finite dimensional real Hilbert space $\mathbb{R}^n$, with some positive definite scalar product (x, y). We shall use $|x|$ for the Hilbert norm of x, such that $|x|^2 = (x, x)$. Since $e^{\frac{i}{2}|x|^2}$ is a bounded continuous function it has a Fourier transform in the sense of tempered distributions and in fact

$$\int e^{\frac{i}{2}|x|^2} e^{i(x,y)} dx = (2\pi i)^{\frac{n}{2}} e^{-\frac{i}{2}|y|^2}, \tag{2.1}$$

with $dx = dx_1 \ldots dx_n$, where $x_i = (e_i, x)$, $e_1, \ldots, e_n$ being some orthonormal base in $\mathbb{R}^n$ with respect to the inner product $(,)$. For a function f in the Schwartz space $\mathcal{S}(\mathbb{R}^n)$ we shall introduce for convenience the notation

$$\widetilde{\int} f(x)dx = (2\pi i)^{-\frac{n}{2}} \int f(x)dx, \tag{2.2}$$

so that $\widetilde{\int} f(x)dx$ is proportional to the usual integral with a normalization factor that depends on the dimension. We get from (2.1) that, for any $\varphi \in \mathcal{S}(\mathbb{R}^n)$

$$\widetilde{\int} e^{\frac{i}{2}|x|^2} \hat{\varphi}(x)dx = \int e^{-\frac{i}{2}|x|^2} \varphi(x)dx, \tag{2.3}$$

where $\hat{\varphi}(x) = \int e^{i(x,y)} \varphi(y)dy$.

Let now $f(x)$ be the Fourier transform of a bounded complex measure μ, $\|\mu\| < \infty$, $f(x) = \int e^{i(x,y)} d\mu(y)$. We shall denote by $\mathcal{F}(\mathbb{R}^n)$ the linear space of functions which are Fourier transforms of bounded complex measures. Since the space of bounded complex measures $\mathcal{M}(\mathbb{R}^n)$ is a Banach algebra under convolution in the total variation norm $\|\mu\|$, we get that $\mathcal{F}(\mathbb{R}^n)$ is a Banach algebra under multiplication in the norm $\|f\|_0 = \|\mu\|$ for $f(x) = \int e^{i(x,y)} d\mu(y)$. The elements in $\mathcal{F}(\mathbb{R}^n)$ are bounded continuous functions and we have obviously $\|f\|_\infty \leq \|f\|_0$. For any $f \in \mathcal{F}(\mathbb{R}^n)$ of the form

$$f(x) = \int e^{i(x,y)} d\mu(y)$$

we define

$$\widetilde{\int} e^{\frac{i}{2}|x|^2} f(x) dx = \int e^{-\frac{i}{2}|x|^2} d\mu(x). \tag{2.4}$$

The right hand side is well defined since $e^{-\frac{i}{2}|x|^2}$ is bounded continuously and $\mu(x)$ is a bounded complex measure. (2.4) defines $\widetilde{\int} e^{\frac{i}{2}|x|^2} f(x) dx$ for all $f \in \mathcal{F}(\mathbb{R}^n)$, and it follows from (2.3) that, for $f \in \mathcal{S}(\mathbb{R}^n)$, $(2\pi i)^{\frac{n}{2}} \widetilde{\int} e^{\frac{i}{2}|x|} f(x) dx$ is just the usual translation invariant integral $\int e^{\frac{i}{2}|x|^2} f(x) dx$ in $\mathbb{R}^n$. Hence $(2\pi i)^{\frac{n}{2}} \widetilde{\int}$ is an extension of the usual translation invariant integral on smooth functions. This extension is a different direction than Lebesgue's extension, namely to the linear space of functions of the form $e^{\frac{i}{2}|x|^2} f(x)$ with $f \in \mathcal{F}(\mathbb{R}^n)$. It follows from (2.4) that this extension is continuous in the sense that

$$\left| \widetilde{\int} e^{\frac{i}{2}|x|^2} f(x) dx \right| \leq \|f\|_0. \tag{2.5}$$

Since $\widetilde{\int}$ is a continuous extension of the normalized integral from $\mathcal{S}(\mathbb{R}^n)$ to $\mathcal{F}(\mathbb{R}^n)$, we shall say that $\widetilde{\int} e^{\frac{i}{2}|x|^2} f(x) dx$ is the normalized integral of the function $e^{\frac{i}{2}|x|^2} f(x)$. By (2.5) we have that

$$\mathcal{F}(f) = \widetilde{\int} e^{\frac{i}{2}|x|^2} f(x) dx \tag{2.6}$$

is a bounded continuous functional on $\mathcal{F}(\mathbb{R}^n)$. We shall call $\mathcal{F}(f)$ the Fresnel integral of f and we shall say that f is a Fresnel integrable function if $f \in \mathcal{F}(\mathbb{R}^n)$.

Remark 1. We have chosen the denomination Fresnel integral for the continuous linear functional (2.6) because of the so called Fresnel integrals in the optical theory of wave diffraction, which are integrals of the form

$$\int_0^W e^{\frac{i\pi}{2} y^2} dy .$$

We now summarize the properties of the Fresnel integral which we have proved above.

Proposition 2.1. *The space $\mathcal{F}(\mathbb{R}^n)$ of Fresnel integrable functions is a Banach-function-algebra in the norm $\|f\|_0$. The Fresnel integral*

$$\mathcal{F}(f) = \widetilde{\int} e^{\frac{i}{2}|x|^2} f(x) dx$$

is a continuous bounded linear functional on $\mathcal{F}(\mathbb{R}^n)$ such that $|\mathcal{F}(f)| \leq \|f\|_0$ and normalized such that $\mathcal{F}(1) = 1$. For any $f(x) \in \mathcal{S}(\mathbb{R}^n)$

$$\overset{\sim}{\int} e^{\frac{i}{2}|x|^2} f(x)\mathrm{d}x = (2\pi i)^{-\frac{n}{2}} \int e^{\frac{i}{2}|x|^2} f(x)\mathrm{d}x.$$

It follows from the fact that $\mathcal{F}(\mathbb{R}^n)$ is a Banach algebra that $\mathcal{F}(f_1, \ldots, f_n)$ is a continuous k-linear form on $\mathcal{F}(\mathbb{R}^n) \times \ldots \times \mathcal{F}(\mathbb{R}^n)$ such that

$$\mathcal{F}(f_1, \ldots, f_k) \leq \|f_1\|_0 \cdots \|f_k\|_0.$$

Moreover sums and products of Fresnel integrable functions are again Fresnel integrable, and the composition with entire functions is also Fresnel integrable.

We shall now define the normalized integral and the Fresnel integral in a real separable Hilbert space. So let $\mathcal{H}$ be a real separable Hilbert space with inner product (x, y) and norm $|x|$. $\mathcal{H}$ is then a separable metric group under addition. Let $\mathcal{M}(\mathcal{H})$ be the Banach space of bounded complex Borel-measures on $\mathcal{H}$. Let μ and ν be two elements in $\mathcal{M}(\mathcal{H})$. The convolution $\mu * \nu$ is defined by

$$\mu * \nu(A) = \int \mu(A - x)\mathrm{d}\nu(x) \tag{2.7}$$

where A is a Borel set. It follows from the fact that $\mathcal{H}$ is a separable metric group that (2.7) is well defined. Moreover $\mathcal{M}(\mathcal{H})$ is in fact a topological semigroup under convolution in the weak topology. For a proof of this fact see [46], Theorem 1.1 p. 57. So that $\mu * \nu$ is simultaneously weakly continuous in μ and ν. From (2.7) we have, for any bounded continuous function f on $\mathcal{H}$, that

$$\int f(x)\mathrm{d}(\mu * \nu)(x) = \int f(x + y)\,\mathrm{d}\mu(x)\mathrm{d}\nu(y),$$

which gives that $\mu * \nu = \nu * \mu$ and $\|\mu * \nu\| \leq \|\mu\| \|\nu\|$, so that $\mathcal{M}(\mathcal{H})$ is a commutative Banach algebra under convolution. We define $\mathcal{F}(\mathcal{H})$ as the space of bounded continuous functions on $\mathcal{H}$ of the form

$$f(x) = \int e^{i(x,y)}\mathrm{d}\mu(y), \tag{2.8}$$

for some $\mu \in \mathcal{M}(\mathcal{H})$. The mapping $\mu \to f$ given by (2.8) is linear and also one-to-one. This is so because, if $f = 0$, then the μ-measure of any set of the form $\{y; (x, y) \geq \alpha\}$ is zero, from which it follows that the μ-measure of any closed convex set is zero. Therefore the μ-measure of any ball is zero, from which it follows that the μ-measure of any strongly measurable set is zero, which implies $\mu = 0$. Therefore introducing the norm $\|f\|_0 = \|\mu\|$, we get that $\mu \to f$ is an isometry onto. It follows from (2.8) that convolution goes into product, so that $\mathcal{F}(\mathcal{H})$, with the norm $\|f\|_0$ is a Banach-function-algebra of continuous bounded functions. From (2.8) we also get that $\|f\|_\infty \leq \|f\|_0$. We now define the normalized integral on $\mathcal{H}$ by

$$\overset{\sim}{\int} e^{\frac{i}{2}|x|^2} f(x)\mathrm{d}x = \int e^{-\frac{i}{2}|x|^2}\mathrm{d}\mu(x), \tag{2.9}$$

for f given by (2.8). We also introduce the Fresnel integral of $f \in \mathcal{F}(\mathcal{H})$ by

$$\mathcal{F}(f) = \int \widetilde{} e^{\frac{i}{2}|x|^2} f(x) \mathrm{d}x. \tag{2.10}$$

We shall call $\mathcal{F}(\mathcal{H})$ the space of Fresnel integrable functions on $\mathcal{H}$.

Proposition 2.2. *The space $\mathcal{F}(\mathcal{H})$ of Fresnel integrable functions is a Banach-function-algebra in the norm $\|f\|_0$. The Fresnel integral $\mathcal{F}(f)$ is a continuous bounded linear functional on $\mathcal{F}(\mathcal{H})$ such that $|\mathcal{F}(f)| \leq \|f\|_0$ and normalized so that $\mathcal{F}(1) = 1$. It follows from the fact that $\mathcal{F}(\mathcal{H})$ is a Banach-algebra that sum and products of Fresnel integrable functions are again Fresnel integrable functions, and so are also compositions with entire functions.*

If $f \in \mathcal{F}(\mathcal{H})$ is a finitely based function, i.e. there exists a finite, dimensional orthogonal projection P in $\mathcal{H}$ such that $f(x) = f(Px)$ or all $x \in \mathcal{H}$, then

$$\int \widetilde{} e^{\frac{i}{2}|x|^2} f(x) \mathrm{d}x = \int_{P\mathcal{H}} \widetilde{} e^{\frac{i}{2}|x|^2} f(x) \mathrm{d}x,$$

where the normalized integral on the right hand side is the normalized integral on the finite dimensional Hilbert space $P\mathcal{H}$ defined previously. This could also be written

$$\mathcal{F}_{\mathcal{H}}(f) = \mathcal{F}_{P\mathcal{H}}(f).$$

Proof. The first part is proved as in Proposition 2.1. To prove the second part we use the definition

$$\int_{\mathcal{H}} \widetilde{} e^{\frac{i}{2}|x|^2} f(x) \mathrm{d}x = \int_{\mathcal{H}} e^{-\frac{i}{2}|x|^2} \mathrm{d}\mu(x) .$$

Now for $f(x)$ to be finitely based with base $P\mathcal{H}$ implies easily that μ has support contained in $P\mathcal{H}$, so that

$$\int_{\mathcal{H}} e^{-\frac{i}{2}|x|^2} \mathrm{d}\mu(x) = \int_{P\mathcal{H}} e^{-\frac{i}{2}|x|^2} \mathrm{d}\mu(x)$$

$$= \int_{P\mathcal{H}} \widetilde{} e^{\frac{i}{2}|x|^2} f(x) \mathrm{d}x .$$

This then proves Proposition 2.2.   □

It follows from its definition that $\mathcal{F}(\mathcal{H})$ is invariant under translations and orthogonal transformations of $\mathcal{H}$, and in fact the induced transformations in $\mathcal{F}(\mathcal{H})$ are isometries of $\mathcal{F}(\mathcal{H})$. It follows from the definition (2.9) of the normalized integral that it is invariant under orthogonal transformations of

$\mathcal{H}$. Consider now a translation $x \to x + a$ of $\mathcal{H}$. The induced transformation in $\mathcal{F}(H)$ is $f \to f_a$, where $f_a(x) = f(x + a)$. If

$$f(x) = \int e^{i(x,y)} \, d\mu(y)$$

then

$$f_a(x) = \int e^{i(x,y)} \, e^{i(a,y)} d\mu(y) \ .$$

By the definition (2.9) we then have

$$\widetilde{\int} e^{\frac{i}{2}|x|^2} f_a(x) dx \ = \ \int e^{-\frac{i}{2}|x|^2} e^{i(a,x)} d\mu(x)$$

$$= e^{\frac{i}{2}|a|^2} \int e^{-\frac{i}{2}|x-a|^2} d\mu(x)$$

$$= e^{\frac{i}{2}|a|^2} \int e^{-\frac{i}{2}|x|^2} d\mu(x+a) \ .$$

Now

$$\int e^{i(x,y)} d\mu(y+a) = e^{-i(x,a)} \int e^{i(x,y)} d\mu(y) = e^{-i(x,a)} f(x) \ .$$

From the relation

$$e^{-i(x,a)} f(x) = \int e^{i(x,y)} d\mu(y+a)$$

we get by (2.9)

$$\widetilde{\int} e^{\frac{i}{2}|x|^2} e^{-i(x,a)} f(x) dx = \int e^{-\frac{i}{2}|x|^2} d\mu(x+a) \ .$$

Hence

$$\widetilde{\int} e^{\frac{i}{2}|x-a|^2} f(x) dx \ = e^{\frac{i}{2}|a|^2} \int e^{-\frac{i}{2}|x|^2} d\mu(x+a)$$

$$= \widetilde{\int} e^{\frac{i}{2}|x|^2} f_a(x) dx \ ,$$

which proves that the normalized integral is also invariant under translations of $\mathcal{H}$. We state these results in the following proposition.

Proposition 2.3. *Let the group of Euclidean transformations $E(\mathcal{H})$ be the group of transformations $x \to Ox + a$, where $a \in \mathcal{H}$ and O is an orthogonal transformation of $\mathcal{H}$ onto $\mathcal{H}$. Then the space of Fresnel integrable functions $\mathcal{F}(\mathcal{H})$ is invariant under $E(\mathcal{H})$, and $E(\mathcal{H})$ is in fact a group of isometries of $\mathcal{F}(\mathcal{H})$. Moreover the normalized integral is invariant under the transformations in $E(\mathcal{H})$.*

We shall now prove the analogue of the Fubini theorem for the normalized integral. So let $\mathcal{H} = \mathcal{H}_1 \oplus \mathcal{H}_2$, then $f \in \mathcal{F}(\mathcal{H})$ defines a continuous function $f(x_1, x_2)$ on $\mathcal{H}_1 \times \mathcal{H}_2$ by $f(x_1 \oplus x_2) = f(x_1, x_2)$. For fixed $x_2 \in \mathcal{H}_2$, $f(x_1, x_2) \in \mathcal{F}(\mathcal{H}_1)$. To see this, we note that we have by definition

$$f(x) = \int e^{i(x,y)} d\mu(y) .$$

Since $||x_1 \oplus x_2||^2 = ||x_1||^2 + ||x_2||^2$ we get that $\mathcal{H}$ and $\mathcal{H}_1 \times \mathcal{H}_2$ are equivalent metric spaces, so that $d\mu(y)$ is actually a measure on the product space $\mathcal{H}_1 \times \mathcal{H}_2$ with the product measure structure. We shall write this measure on $\mathcal{H}_1 \times \mathcal{H}_2$ as $d\mu(y_1, y_2)$.

Hence

$$f(x_1, x_2) = \int e^{i(x_1, y_1)} e^{i(x_2, y_2)} d\mu(y_1, y_2) . \tag{2.11}$$

Consider now the measure $\mu_{x_2} \in \mathcal{M}(\mathcal{H}_1)$ defined by

$$\int_{\mathcal{H}_1} \varphi(y_1) d\mu_{x_2}(y_1) = \int \varphi(y_1) e^{i(x_2, y_2)} d\mu(y_1, y_2) . \tag{2.12}$$

By the usual Fubini theorem we then have that

$$f(x_1, x_2) = \int_{\mathcal{H}_1} e^{i(x_1, y_1)} d\mu_{x_2}(y_1) . \tag{2.13}$$

This proves that, for fixed $x_2 \in \mathcal{H}_2$, $f(x_1, x_2) \in \mathcal{F}(\mathcal{H}_1)$. Hence the normalized integral

$$g(x_2) = \int_{\mathcal{H}_1}^{\sim} e^{\frac{i}{2}|x_1|^2} f(x_1, x_2) dx_1 \tag{2.14}$$

is well defined. We shall now see that $g(x_2) \in \mathcal{F}(\mathcal{H}_2)$. By the definition of the normalized integral and (2.13) we have that

$$\int_{\mathcal{H}_1}^{\sim} e^{\frac{i}{2}|x_1|^2} f(x_1, x_2) dx_1 = \int_{\mathcal{H}_1} e^{-\frac{i}{2}|y_1|^2} d\mu_{x_2}(y_1) \tag{2.15}$$

and by (2.12) this is equal to

$$\int e^{-\frac{i}{2}|y_1|^2} e^{i(x_2, y_2)} d\mu(y_1, y_2) . \tag{2.16}$$

Hence

$$g(x_2) = \int_{\mathcal{H}_2} e^{i(x_2, y_2)} d\nu(y_2) , \tag{2.17}$$

where $\nu \in \mathcal{M}(\mathcal{H}_2)$ is defined by

$$\int_{\mathcal{H}_2} \varphi(y_2) \mathrm{d}\nu(y_2) = \int e^{-\frac{1}{2}|y_1|^2} \varphi(y_2) \mathrm{d}\mu(y_1, y_2) \ . \tag{2.18}$$

This then proves that $g(x_2) \in \mathcal{F}(\mathcal{H}_2)$, and the normalized integral

$$\widetilde{\int_{\mathcal{H}_2}} e^{\frac{i}{2}|x_2|^2} g(x_2) \mathrm{d}x_2 \tag{2.19}$$

is well defined.

By (2.17) and the definition of the normalized integral we have that

$$\widetilde{\int_{\mathcal{H}_2}} e^{\frac{i}{2}|x_2|^2} g(x_2) \mathrm{d}x_2 = \int_{\mathcal{H}_2} e^{-\frac{i}{2}|y_2|^2} \mathrm{d}\nu(y_2)$$

which by Fubini and (2.18) is equal to

$$\int e^{-\frac{i}{2}|y_1|^2} e^{-\frac{i}{2}|y_2|^2} \mathrm{d}\mu(y_1, y_2) = \widetilde{\int_{\mathcal{H}}} e^{\frac{i}{2}|x|^2} f(x) \mathrm{d}x \ .$$

We have now proven the following proposition, which we may also call the Fubini theorem for the normalized integral.

Proposition 2.4 (Fubini theorem). *Let $\mathcal{H} = \mathcal{H}_1 \oplus \mathcal{H}_2$ be the orthogonal sum of two subspaces $\mathcal{H}_1$ and $\mathcal{H}_2$. For $f(x) \in \mathcal{F}(\mathcal{H})$ set $f(x_1, x_2) = f(x_1 \oplus x_2)$ with $x_1 \in \mathcal{H}_1$ and $x_2 \in \mathcal{H}_2$. Then for fixed x_2, $f(x_1, x_2)$ is in $\mathcal{F}(\mathcal{H}_1)$ and*

$$g(x_2) = \widetilde{\int_{\mathcal{H}_1}} e^{\frac{i}{2}|x_1|^2} f(x_1, x_2) \mathrm{d}x_1 \tag{2.20}$$

is in $\mathcal{F}(\mathcal{H}_2)$. Moreover

$$\widetilde{\int_{\mathcal{H}_2}} e^{\frac{i}{2}|x_2|^2} g(x_2) \mathrm{d}x_2 = \widetilde{\int_{\mathcal{H}_2}} e^{\frac{i}{2}|x_2|^2} \left(\widetilde{\int_{\mathcal{H}_1}} e^{\frac{i}{2}|x_1|^2} f(x_1, x_2) \mathrm{d}x_1 \right) \mathrm{d}x_2$$

$$= \widetilde{\int_{\mathcal{H}}} e^{\frac{i}{2}|x|^2} f(x) \mathrm{d}x \ .$$

Remark. The normalized integral on a separable Hilbert space is the same as the functional $\mathcal{F}$ defined by Ito in [13]

$$\int^{\sim} e^{\frac{i}{2}|x|^2} f(x)\mathrm{d}x = \mathcal{F}(e^{\frac{i}{2}|x|^2} f) \,.$$

Ito takes a completely different definition, he defines namely $\mathcal{F}$ for $f \in \mathcal{F}(\mathcal{H})$ by

$$\mathcal{F}(e^{\frac{i}{2}|x|^2} f) = \lim_V \prod_{j=1}^{\infty} (1 - i\lambda_j)^{\frac{1}{2}} E(e^{\frac{i}{2}|x|^2} f; a, V) \,,$$

where $E(g; a, V)$ is the expectation of g with respect to the Gaussian measure with mean $a \in \mathcal{H}$ and covariance operator V, where V is a strictly positive definite symmetric trace class operator on $\mathcal{H}$ with eigenvalues λ_j, such that $\sum_{j=1}^{\infty} \lambda_j < \infty$. The limit is taken along the directed system of all strictly positive definite trace class operators with the direction given by the relation $<$, where $V_1 < V_2$ if and only if $V_2 - V_1$ is positive. Ito proves that this limit exists and is independent of a and moreover that it is invariant under nearly isometric transformations, in the sense that

$$F(e^{\frac{i}{2}|Cx|^2} f(Cx)) = J(C)^{-1} F(e^{\frac{i}{2}|x|^2} f) \,,$$

where $Cx = Ax + b$, with $b \in \mathcal{H}$ and A a one-to-one map of $\mathcal{H}$ such that

$$tr([(A^*A)^{\frac{1}{2}} - 1]^{\frac{\alpha}{2}}) < \infty$$

for some $a < 1$. $J(C)$ is defined by $J(C) = \prod_{j=1}^{\infty}(1 + \alpha_j)$, where α_j are the eigenvalues of $(A^*A)^{\frac{1}{2}} - 1$.

Instead of Ito's definition we have used Parseval relation (2.9) as a definition for the normalized integral because we shall later need a generalization of the normalized integral to spaces with indefinite metric. These generalizations will be very natural in our setting and in fact defined again by a sort of Parseval relation. We feel also that the definition of the normalized integral by the Parseval relation (2.9) gives a nice and simple introduction to the properties of the normalized integral.

We also want to point out that the first part of the next section, i.e. the formula for the finite time transition amplitude, was first derived by Ito [13], but we shall give the proof of it here partly for the sake of completeness and also because our proof is independent of Ito's, and it will later on be extended to cover different situations.

Notes

The name "Fresnel integral" is inspired by a particular integral which appeared in the framework of classical optics, namely the "classical Fresnel integrals" $\int_0^w \sin(\frac{\pi}{2}x^2)$ $\mathrm{d}x$, $\int_0^w \cos(\frac{\pi}{2}x^2)\mathrm{d}x$, $\int_0^w e^{i\frac{\pi}{2}x^2}\mathrm{d}x$.

It has been introduced in connection with (rigorous) Feynman path integrals in the first edition of this book.

The study of oscillatory integrals in finite dimensions, of which the above classical Fresnel integrals are particular cases, has been developed in the nineteenth century in work by Stokes, in connection on one hand with the classical method of stationary phase and, on the other hand, with the theory of improper Riemann integrals. The classical method of stationary phase is discussed, e.g., in [226, 145, 387]. Recent developments are connected with the theory of singularities of mappings (including catastrophe theory), see, e.g., [123], and the theory of Fourier integral operators [220, 278, 279, 363, 364, 366, 424].

The concept of Fresnel integrable function and Fubini theorem for oscillatory integrals appeared first in the first edition of this book, however the role of the Banach algebra $\mathcal{F}(\mathcal{H})$ had already been pointed out in the paper by Ito [13] in 1967. For recent theoretical developments on rigorously defined Feynman path integrals see Chapter 10 and, e.g., [214, 274, 280, 281, 298, 299, 303, 301, 305, 316, 321, 388, 416, 418, 422, 439, 445, 446].

3

The Feynman Path Integral in Potential Scattering

In this section we consider the Schrödinger equation for a quantum mechanical particle in $\mathbb{R}^n$ under the influence of a potential $V(x)$. The Schrödinger equation for the wavefunction $\psi(x,t)$ is given by

$$i\hbar\frac{\partial\psi}{\partial t} = -\frac{\hbar^2}{2m}\triangle\psi + V\psi \tag{3.1}$$

where m is the mass of the particle, $\hbar$ is Planck's constant and $\triangle$ is the Laplacian $\triangle\psi = \sum_{i=1}^{n}\frac{\partial^2\psi}{\partial x_i^2}$. For typographical reasons we choose units for time and length such that $\hbar = 1$, in which case we get

$$i\frac{\partial\psi}{\partial t} = -\frac{1}{2m}\triangle\psi + V\psi \ . \tag{3.2}$$

$H_0 = -\frac{1}{2m}\triangle$ is a self adjoint operator in $L_2(\mathbb{R}^n)$ on its natural domain of definition. In what follows V will be a bounded continuous real function on $\mathbb{R}^n$, hence $H = H_0 + V$ is also a self adjoint operator with the same domain as H_0. The solution of the initial value problem for (3.2) is therefore given by

$$\psi(x,t) = (\mathrm{e}^{-itH}\varphi)(x) \tag{3.3}$$

with initial data $\varphi \in L_2(\mathbb{R}^n)$, where e^{-itH} is the unitary group in $L_2(\mathbb{R}^n)$ generated by $-H$. We shall now express (3.3) as a so-called Feynman path integral. In fact (3.3) will be given by the normalized integral of e^{iS}, where S is the classical action, over all path's for the particle ending at x at time t. This expression was suggested by Feynman and proved by Ito [13], and we shall therefore call it the Feynman–Ito formula. It is the correspondent for the Schrödinger equation of the Feynman–Kac formula for the heat equation.

It is well known that e^{-itH} can be expanded in powers of V. We have namely that

$$\frac{d}{dt}e^{itH_0}e^{-itH} = -i\,e^{itH_0}Ve^{-itH}$$

$$= -i\,V(t)e^{itH_0}e^{-itH}\ ,$$

where $V(t) = e^{itH_0}Ve^{-itH_0}$. Integrating this formula we get

$$e^{itH_0}e^{-itH} = 1 - i\int_0^t V(t_1)e^{it_1H_0}e^{-it_1H}dt_1\ . \tag{3.4}$$

By iteration we have then

$$e^{itH_0}e^{-itH} = \sum_{n=0}^{\infty}(-i)^n \int\cdots\int_{t\geq t_1\geq\cdots\geq t_n\geq 0} V(t_1)\ldots V(t_n)dt_1\ldots dt_n\ , \tag{3.5}$$

which by the norm boundedness of V is obviously norm convergent for all t. By substitution under the integrals we get from (3.5)

$$e^{-itH} = \sum_{n=0}^{\infty}(-i)^n \int\cdots\int_{0\leq t_1\cdots\leq t_n\leq 0} e^{-itH_0}V(t_n)\ldots V(t_1)dt_1\ldots dt_n\ , \tag{3.6}$$

or more explicitly

$$e^{-itH} = \sum_{n=0}^{\infty}(-i)^n \int\cdots\int_{0\leq t_1\leq\cdots\leq t_n\leq t} e^{-i(t-t_n)H_0}Ve^{-i(t_n-t_{n-1})H_0}\ldots$$

$$\ldots e^{-i(t_2-t_1)H_0}Ve^{-it_1H_0}dt_1\ldots dt_n\ . \tag{3.7}$$

We shall now assume that V is of the form

$$V(x) = \int_{\mathbb{R}^n} e^{i\alpha x}d\mu(x) \tag{3.8}$$

and that

$$\varphi(x) = \int_{\mathbb{R}^n} e^{i\alpha x}d\nu(\alpha) \tag{3.9}$$

where $\alpha x = \sum_{i=1}^{\infty}\alpha_i x_i$, for some bounded complex measures μ and ν on $\mathbb{R}^n$. Since $e^{i\alpha x}$ is a generalized eigenfunction for H_0, with

$$e^{-itH_0}e^{i\alpha x} = e^{-\frac{it}{2m}\alpha^2}e^{i\alpha x} \tag{3.10}$$

we get by substitution of (3.8), (3.9) and (3.7) in (3.3) that

$$\psi(x,t) = \sum_{n=0}^{\infty} (-i)^n \int_{0 \le t_1 \le \dots \le t_n \le t} \dots \int \int \dots \int$$

$$e^{-\frac{i}{2m} \left[(t-t_n)(\alpha_0 + \dots + \alpha_n)^2 + (t_n - t_{n-1})(\alpha_0 + \dots + \alpha_{n-1})^2 + \dots + (t_2 - t_1)(\alpha_0 + \alpha_1)^2 + t_1 \alpha_0^2 \right]} \cdot$$

$$\cdot \; e^{i\left(\sum_{j=0}^{n} \alpha_j\right) x} d\nu(\alpha_0) \prod_{j=1}^{n} (d\mu(\alpha_j) dt_j) \; . \tag{3.11}$$

Introducing the notation $t_0 = 0$ we may simplify the exponent in (3.11) and we get

$$\psi(x,t) = \sum_{n=0}^{\infty} (-i)^n \int_{0 \le t_1 \le \dots \le t_n \le t} \dots \int \int \dots \int e^{-\frac{i}{2m} \sum_{j,k=0}^{n} (t - t_j \vee t_k) \alpha_j \alpha_k} \; e^{i\left(\sum_{j=0}^{n} \alpha_j\right) x} \cdot$$

$$\cdot \; d\nu(\alpha_0) \prod_{j=1}^{n} (d\mu(\alpha_j) dt_j) \tag{3.12}$$

where $\sigma \vee \tau = \max\{\sigma, \tau\}$.

By the symmetry of the integrand we have then

$$\psi(x,t) = \sum_{n=0}^{\infty} \frac{(-i)^n}{n!} \int_0^t \dots \int \int_0^t \dots \int e^{-\frac{i}{2m} \sum_{j,k=0}^{n} (t - t_j \vee t_k) \alpha_j \alpha_k} \; e^{i\left(\sum_{j=0}^{n} \alpha_j\right) x} \cdot$$

$$\cdot \; d\nu(\alpha_0) \sum_{j=1}^{n} (d\mu(\alpha_j) dt_j) \; . \tag{3.13}$$

Now $G_{ij}(\sigma, \tau) = (t - \sigma \vee \tau)\delta_{ij}$ is the Green's function or the kernel of the inverse operator of $-\frac{d^2}{d\tau^2}$ as a self adjoint operator on $L_2([0,t]; \mathbb{R}^n)$ with boundary conditions $\frac{du}{d\tau}(0) = u(t) = 0$.

Hence if we introduce the real Hilbert space $\mathcal{H}$ of real continuous functions $\gamma(\tau)$ from $[0,t]$ to $\mathbb{R}^n$ such that $\frac{d\gamma}{d\tau} \in L_2([0,t]; \mathbb{R}^n)$ and $\gamma(t) = 0$ with inner product $(\gamma_1, \gamma_2) = m \int_0^t \frac{d\gamma_1}{d\tau} \cdot \frac{d\gamma_2}{d\tau} d\tau$, then $\gamma_{(\sigma,i)}(\tau, j) = G_{ij}(\sigma, \tau) = (t - \sigma \vee \tau)\delta_{ij}$ is in $\mathcal{H}$ for any (σ, i), and for any $\gamma \in \mathcal{H}$ we have

$$\gamma(\sigma, i) = m(\gamma, \gamma_{(\sigma,i)}) \; . \tag{3.14}$$

It follows from (3.14) that the functions on $\mathcal{H}$ given by

$$\varphi(\gamma(0) + x) = \int_{\mathbb{R}^n} e^{i\gamma(0) \cdot \alpha} e^{i\alpha x} d\nu(\alpha)$$

and

$$\int_0^t V(\gamma(\tau) + x)\mathrm{d}t = \int_0^t \int_{\mathbb{R}^n} e^{i\gamma(\tau)\cdot\alpha} e^{ix\alpha} \mathrm{d}\mu(\alpha)\mathrm{d}t$$

both are Fourier transforms of bounded measures on $\mathcal{H}$ so that both functions are in $\mathcal{F}(\mathcal{H})$, where $\mathcal{F}(\mathcal{H})$ is the space of Fresnel integrable functions on $\mathcal{H}$ defined in the previous section. By Proposition 2.2 of Chap. 2 we therefore have that the continuous function $f(\gamma)$ on $\mathcal{H}$ given by

$$f(\gamma) = e^{-i\int_0^t V(\gamma(\tau)+x)\mathrm{d}t} \varphi(\gamma(0) + x) \tag{3.15}$$

is in $\mathcal{F}(\mathcal{H})$. Hence the normalized integral $\int_{\mathcal{H}}^{\sim} e^{\frac{i}{2}|\gamma|^2} f(\gamma)\mathrm{d}\gamma$ is well defined. Since the exponent is invariant under the transformation $\gamma \to \gamma - x$, it is natural to introduce also the following two notations for this integral:

$$\int_{\mathcal{H}}^{\sim} e^{\frac{i}{2}|\gamma|^2} f(\gamma)\mathrm{d}\gamma \equiv \int_{\gamma(t)=0}^{\sim} e^{\frac{i}{2}m\int_0^t |\frac{\mathrm{d}\gamma}{\mathrm{d}\tau}|^2 \mathrm{d}\tau} f(\gamma)\mathrm{d}\gamma \tag{3.16}$$

$$\equiv \int_{\gamma(t)=x}^{\sim} e^{\frac{i}{2}m\int_0^t |\frac{\mathrm{d}\gamma}{\mathrm{d}\tau}|^2 \mathrm{d}\tau} f(\gamma - x)\mathrm{d}\gamma \, , \tag{3.17}$$

where $(\gamma - x)(\tau) = \gamma(t) - x$.

We shall now compute the normalized integral (3.16), with $f(\gamma)$ given by (3.15). By Proposition 2.2 of Chap. 2 we have that

$$\int_{\mathcal{H}}^{\sim} e^{\frac{i}{2}|\gamma|^2} f(\gamma)\mathrm{d}\gamma = \sum_{n=0}^{\infty} \frac{(-i)^n}{n!} \int_{\mathcal{H}}^{\sim} e^{\frac{i}{2}|\gamma|^2} \left(\int_0^t V(\gamma(\tau) + x)\mathrm{d}t \right)^n \varphi(\gamma(0) + x)\mathrm{d}\gamma$$

$$= \sum_{n=0}^{\infty} \frac{(-i)^n}{n!} \int_{\mathcal{H}}^{\sim} e^{\frac{i}{2}|\gamma|^2} \Bigg[\int_0^t \cdots \int_0^t \int \cdots \int e^{i\left(\sum_{j=1}^n \gamma(t_j)\alpha_j + \gamma(0)\alpha_0\right)}$$

$$e^{ix\alpha_0}\mathrm{d}\nu(\alpha_0) \prod_{j=1}^n e^{ix\alpha_j}\mathrm{d}\mu(\alpha_j)\mathrm{d}t_j \Bigg]\mathrm{d}\gamma \, . \tag{3.18}$$

Now, by expressing $\sum_{j=1}^n \gamma(t_j)\alpha_j + \gamma(0)\alpha_0$ as a scalar product in $\mathcal{H}$ by the formula (3.14), we see that the nth term in the sum in (3.18) is the normalized integral of the Fourier transform of a bounded measure on $\mathcal{H}$. Hence by the definition (2.9) of the normalized integral we get that (3.18) is equal to

$$\sum_{n=0}^{\infty} \frac{(-i)^n}{n!} \int_0^t \cdots \int_0^t \int \cdots \int e^{-\frac{i}{2m}\sum_{j,k=0}^n \alpha_j G(t_j,t_k)\alpha_k}$$

$$e^{i\left(\sum_{j=0}^n \alpha_j\right)x}\mathrm{d}\nu(\alpha_0) \prod_{j=1}^n \mathrm{d}\mu(\alpha_j)\mathrm{d}t_j \, , \tag{3.19}$$

where we have introduced the notation to $t_0 = 0$. By the definition of the matrix $G(\sigma, \tau)$ we see that (3.19) is equal to (3.13). Hence we have proved the Feynman–Ito formula, namely using the notation (3.17):

$$\psi(x,t) = \int\limits_{\gamma(t)=x}^{\sim} e^{\frac{im}{2}\int_0^t |\frac{d\gamma}{d\tau}|^2 d\tau} \cdot e^{-i\int_0^t V(\gamma(\tau))d\tau} \varphi(\gamma(0))d\gamma . \qquad (3.20)$$

Introducing the classical action along the path γ in the time interval $[0,t]$

$$S_t(\gamma) = \int\limits_0^t \frac{m}{2}\left|\frac{d\gamma}{d\tau}\right|^2 d\tau - \int\limits_0^t V(\gamma(\tau))d\tau ,$$

(3.20) may be written in more compact notations as

$$\psi(x,t) = \int\limits_{\gamma(t)=x}^{\sim} e^{iS_t(\gamma)}\varphi(\gamma(0))d\gamma . \qquad (3.21)$$

With units such that $\hbar \neq 1$ we easily get the formula

$$\psi(x,t) = \int\limits_{\gamma(t)=x}^{\sim} e^{i/\hbar\, S_t(\gamma)}\varphi(\gamma(0))d\gamma \qquad (3.22)$$

for the solution of the Schrödinger equation (3.1). We formulate this result, which was first established by Ito, in the following theorem.

Theorem 3.1 (The Feynman–Ito formula).
Let V and φ be Fourier transforms of bounded complex measures in $\mathbb{R}^n$. Let $\mathcal{H}$ be the real Hilbert space of continuous paths γ from $[0,t]$ to $\mathbb{R}^n$ such that $\gamma(t) = 0$ and $\frac{d\gamma}{d\tau} \in L_2([0,t]; \mathbb{R}^n)$ with inner product $(\gamma_1, \gamma_2) = \frac{m}{\hbar}\int_0^t \left(\frac{d\gamma_1}{d\tau} \cdot \frac{d\gamma_2}{d\tau}\right) d\tau$. Then

$$f(\gamma) = e^{-\frac{i}{\hbar}\int_0^t V(\gamma(\tau)+x)d\tau}\varphi(\gamma(0) + x)$$

is in $\mathcal{F}(\mathcal{H})$, the space of Fresnel integrable functions on $\mathcal{H}$ and the solution of the Schrödinger equation

$$i\hbar\frac{\partial\psi}{\partial t} = -\frac{\hbar^2}{2m}\Delta\psi + V\psi$$

with boundary condition $\psi(x,0) = \varphi(x)$ is given by the normalized integral

$$\psi(x,t) = \int\limits_{\mathcal{H}}^{\sim} e^{\frac{i}{2}|\gamma|^2}f(\gamma)d\gamma ,$$

i.e.

$$\psi(x,t) = \int\limits_{\gamma(t)=x} e^{\frac{im}{2\hbar} \int\limits_0^t |\frac{d\gamma}{d\tau}|^2 d\tau} \cdot e^{-\frac{i}{\hbar} \int\limits_0^t V(\gamma(\tau))d\tau} \varphi(\gamma(0))d\gamma \ .$$

$\square$ 1,2

We shall now proceed to study the wave operator and the scattering operator in terms of Feynman path integrals. Let us again use units where $\hbar = 1$. Let us also, for typographically simplicity, assume that $m = 1$.

The wave operators $W_{\pm}$ are defined by

$$W_{\pm} = \underset{t \to \pm\infty}{st. \lim} \, e^{-itH} e^{itH_0} \ , \tag{3.23}$$

whenever these limits exist.[3] It is well known that these limits exist for a wide range of potentials $V(x)$ that fall off sufficiently fast. In what follows we shall assume that $V(x)$ is such that the limits (3.23) exist. So that

$$V(x) = \int\limits_{\mathbb{R}^n} e^{i\alpha x} d\mu(\alpha)$$

---

[1] By the "translation invariance" of the normalized integral we have

$$\int\limits_{\mathcal{H}} e^{\frac{i}{2\hbar}|\gamma|^2} f(\gamma)d\gamma = \int\limits_{\mathcal{H}} e^{\frac{i}{2\hbar}|\gamma+\gamma_0|^2} f(\gamma+\gamma_0)d\gamma \ ,$$

where γ_0 is any element of $\mathcal{H}$. Hence

$$\int\limits_{\gamma(t)=x} e^{\frac{im}{2\hbar} \int\limits_0^t |\frac{d\gamma}{d\tau}|^2 d\tau} e^{-\frac{i}{\hbar} \int\limits_0^t V(\gamma(\tau))d\tau} \varphi(\gamma(0))d\gamma \ ,$$

although defined by replacing in its argument $\gamma(\tau)$ by $\gamma(\tau) + x$, and interpreting the result as $\int_{\mathcal{H}} e^{\frac{i}{2}|\gamma|^2} f(\gamma)d\gamma$, is actually independent of the chosen particular translation, which could be replaced by $x + \gamma_0$, with $\gamma_0 \in \mathcal{H}$ arbitrary, and only depends on its value for $\tau = $ t, which is required to be x.

[2] The above Theorem 3.1 has been stated for the Schrödinger equation of a particle moving in $\mathbb{R}^n$ under the action of the potential V. From the proof it is evident that the same results hold for a system of N non relativistic quantum mechanical particles moving each in d dimensional space, thus with $n = Nd$, under the influence of a potential V which is only restricted to belong to $\mathcal{F}(\mathbb{R}^n)$ but is otherwise arbitrary, hence can be e.g. a sum of ν-body translation invariant potentials ($\nu = 1, 2, ...$). In this case $-\frac{\hbar^2}{2m}\triangle$ is replaced by $-\frac{\hbar^2}{2}\sum_{i=1}^n \frac{1}{m_i}\frac{\partial^2}{\partial x_i^2}$ and (γ_1, γ_2) by $(\gamma_1, \gamma_2) = \frac{1}{\hbar}\int_0^t \frac{d\gamma_1}{d\tau} \cdot M \frac{d\gamma_2}{d\tau} d\tau$, where M is the matrix $(M)_{ij} = m_i\delta_{ij}, i, j = 1, \ldots, n$.

[3] The elementary definitions of mathematical scattering theory are e.g. in [47]. For recent work see e.g. [32].

and for instance[4]

$$|V(x)| \le C(1+|x|)^{-1-\epsilon}$$

(see for example [43]).

By expanding $\mathrm{e}^{-\mathrm{i}sH_0}\mathrm{e}^{-\mathrm{i}(t-s)H}\mathrm{e}^{-\mathrm{i}tH_0}$ in powers of V in a similar manner as in (3.5) we get, with $V(t) = \mathrm{e}^{-\mathrm{i}tH_0}V\mathrm{e}^{\mathrm{i}tH_0}$ that

$$\mathrm{e}^{-\mathrm{i}sH_0}\mathrm{e}^{-\mathrm{i}(t-s)H}\mathrm{e}^{-\mathrm{i}tH_0} = \sum_{n=0}^{\infty}(-\mathrm{i})^n \int\cdots\int_{s\le t_1\le\cdots\le t_n\le t} V(t_1)\ldots V(t_n)\mathrm{d}t_1\ldots\mathrm{d}t_n\ ,$$

where the sum is norm convergent for all s and t, i.e.

$$\mathrm{e}^{-\mathrm{i}sH_0}\mathrm{e}^{-\mathrm{i}(t-s)H}\mathrm{e}^{-\mathrm{i}tH_0} = \sum_{n=0}^{\infty}(-1)^n \int_{s\le t_1\le}\cdots\int_{\le t_n\le t} \mathrm{e}^{-\mathrm{i}t_1 H_0}V\mathrm{e}^{-\mathrm{i}(t_2-t_1)H_0}\ldots$$
$$\ldots\mathrm{e}^{-\mathrm{i}(t_n-t_{n-1})H_0}V\mathrm{e}^{\mathrm{i}t_n H_0}\mathrm{d}t_1\ldots\mathrm{d}t_n\ .$$

With

$$\varphi(x) = \int \mathrm{e}^{\mathrm{i}\beta x}\mathrm{d}\nu(\beta)$$

we get then

$$\left(\mathrm{e}^{-\mathrm{i}sH_0}\mathrm{e}^{-\mathrm{i}(t-s)H}\mathrm{e}^{\mathrm{i}tH_0}\varphi\right)(x) = \sum_{n=0}^{\infty}(-\mathrm{i})^n \int\cdots\int\int\cdots\int_{s\le t_1\le\cdots\le t_n\le t}$$
$$\mathrm{e}^{-\frac{\mathrm{i}}{2}\left[t_1(\alpha_1+\cdots+\alpha_n+\beta)^2+(t_2-t_1)(\alpha_2+\cdots+\beta)^2+\cdots+(t_n-t_{n-1})(\alpha_n+\beta)^2-t_n\beta^2\right]}$$
$$\cdot\ \mathrm{e}^{\mathrm{i}\left(\sum_{j=1}^{n}\alpha_j+\beta\right)x}\mathrm{d}\nu(\beta)\prod_{i=1}^{n}(\mathrm{d}\mu(\alpha_i)\mathrm{d}t_i)\ . \tag{3.24}$$

We have

$$t_1(\alpha_1+\cdots+\alpha_n+\beta)^2 + (t_2-t_1)(\alpha_2+\cdots+\beta)^2$$
$$+\cdots+(t_n-t_{n-1})(\alpha_n+\beta)^2 - t_n\beta^2 \tag{3.25}$$
$$= t_1\alpha_1^2 + 2t_1\alpha_1(\alpha_2+\cdots+\alpha_n+\beta) + t_2\alpha_2^2 + 2t_2\alpha_2(\alpha_3+\ldots\alpha_n+\beta)$$
$$+\cdots+t_n\alpha_n^2 + 2t_n\alpha_n\beta$$
$$= \sum_{ij=1}^{n} t_i\wedge t_j\alpha_i\alpha_j + 2\beta\sum_{i=1}^{n}t_i\alpha_i\ , \tag{3.26}$$

where $s\wedge t = \min\{s,t\}$.

---

[4] This assumption is actually enough for proving the completeness of the wave operators, in the sense that Range W_+ = Range W_-, see [48]. A slightly weaker condition, sufficient for the existence of the wave operators, is ([49]):

$$\int |V(x)|^2(1+|x|)^{2-n+\epsilon}\mathrm{d}x < \infty,\quad \epsilon > 0\ .$$

See also [11], 3).

If we introduce $\delta = -\sum_{i=1}^{n} \alpha_i - \beta$, (3.25) may also be written

$$t_1\delta^2 + (t_2 - t_1)(\delta + \alpha_1)^2 + \cdots + (t_n - t_{n-1})(\delta + \alpha_1 + \cdots + \alpha_{n-1})^2$$
$$-t_n(\delta + \alpha_1 + \cdots + \alpha_n)^2$$
$$= -t_1\alpha_1^2 - 2t_1\delta\alpha_1 - t_2\alpha_2^2 - 2t_2\alpha_2(\delta + \alpha_1)$$
$$+ \cdots - t_n\alpha_n^2 - 2t_n\alpha_n(\delta + \alpha_1 + \cdots + \alpha_{n-1})$$
$$= -\sum_{ij} t_i \vee t_j \alpha_i \alpha_j - 2\delta \sum_{i=1}^{n} t_i\alpha_i , \tag{3.27}$$

where $s \vee t = \max\{s, t\}$.

By the identity of (3.25) with (3.26) and (3.27) and the fact that $|s - t| = s \vee t - s \wedge t$ we have that (3.25) is also equal to

$$\sum_{ij=1}^{n} -\frac{1}{2}|t_i - t_j|\alpha_i\alpha_j + (\beta - \delta)\sum_{i=1}^{n} t_i\alpha_i . \tag{3.28}$$

If $s = 0$ and $t > 0$ we get from (3.24) and (3.25) that

$$\left(e^{-itH}e^{itH_0}\varphi\right)(x) = \sum_{n=0}^{\infty} (-i)^n \int \cdots \int_{0 \le t_1 \le \ldots \le t_n \le t} \int \int \cdots \int$$
$$\cdot\ e^{-\frac{i}{2}\sum_{jk=1}^{n} t_j \wedge t_k \alpha_j \alpha_k - i\beta \sum_{j=1}^{n} t_j\alpha_j + i\left(\sum_{j=1}^{n} \alpha_j + \beta\right)x}$$
$$\cdot\ d\nu(\beta) \prod_{j=1}^{n} d\mu(\alpha_j) dt_j . \tag{3.29}$$

By the substitution $t_i \to -t_i, i = 1, \ldots, n$ we get that (3.29) is equal to

$$\sum_{n=0}^{\infty} (-i)^n \int \cdots \int_{-t \le t_n \le \ldots \le t_1 \le 0} \int \int \cdots \int e^{-\frac{i}{2}\sum_{jk=1}^{n} -t_j \vee t_k \alpha_j \alpha_k + i\beta \sum_{j=1}^{n} t_j\alpha_j + i\left(\sum_{j=1}^{n} \alpha_j + \beta\right)x}$$
$$d\nu(\beta) \prod_{j=1}^{n} d\mu(\alpha_j) dt_j ,$$

which, by the symmetry of the integrand, gives

$$\left(e^{-itH}e^{itH_0}\varphi\right)(x) = \sum_{n=0}^{\infty} \frac{(-i)^n}{n!} \int_{-t}^{0} \cdots \int_{-t}^{0} \int \int \cdots \int$$
$$\cdot\ e^{-\frac{i}{2}\sum_{jk=1}^{n} -t_j \vee t_k \alpha_j \alpha_k + i\beta \sum_{j=1}^{n} t_j\alpha_j + i\left(\sum_{j=1}^{n} \alpha_j + \beta\right)x}$$
$$\cdot\ d\nu(\beta) \prod_{j=1}^{n} d\mu(\alpha_j) dt_j . \tag{3.30}$$

Consider now the separable real Hilbert space $\mathcal{H}_-$ of continuous functions γ from $[-\infty, 0]$ to $\mathbb{R}^n$ such that $\gamma(0) = 0$ and $\frac{d\gamma}{d\tau}$ is in $L_2([-\infty, 0]; \mathbb{R}^n)$ with norm given by

$$|\gamma|^2 = \int\limits_{-\infty}^{0} \left|\frac{d\gamma}{d\tau}\right|^2 d\tau . \tag{3.31}$$

We have that $\gamma_{(s,i)}(t,j) = -s \vee t \cdot \delta_{ij}$ is in $\mathcal{H}_-$ and for all $\gamma \in \mathcal{H}_-$ we have

$$(\gamma, \gamma_{s,i}) = \gamma(s,i) . \tag{3.32}$$

From this it follows that

$$f(\gamma) = \int\limits_{-t}^{0} V(\gamma(\tau) + \beta\tau + x)d\tau = \int\limits_{-t}^{0} \int e^{i\alpha\gamma(\tau)} \cdot e^{i\alpha(\beta\tau+x)}d\mu(\alpha)d\tau \tag{3.33}$$

is in $\mathcal{F}(\mathcal{H}_-)$. Hence, by Proposition 2.2 of Chap. 2, $e^{-if(\gamma)}$ again in $\mathcal{F}(\mathcal{H}_-)$ and the normalized integral

$$\int\limits_{\mathcal{H}_-}^{\sim} e^{\frac{i}{2}|\gamma|^2} e^{-if(\gamma)}d\gamma \tag{3.34}$$

is well defined. By using (3.32) we may compute explicitly the normalized integral (3.34) in the same way as in the proof of the Theorem 3.1 and we get

$$\int\limits_{\mathcal{H}_-}^{\sim} e^{\frac{i}{2}|\gamma|^2} e^{-if(\gamma)}d\gamma = \sum_{n=0}^{\infty} \frac{(-i)^n}{n!} \tag{3.35}$$

$$\int\limits_{-t}^{0} \cdots \int\limits_{-t}^{0} \int \cdots \int e^{-\frac{1}{2}\sum\limits_{jk=1}^{n} -t_j \vee t_k \alpha_j \alpha_k + i\beta \sum\limits_{j=1}^{n} t_j \alpha_j + \left(i\sum\limits_{j=1}^{n} \alpha_j\right)x} \prod_{j=1}^{n} d\mu(\alpha_j)dt_j .$$

Introducing now the notation

$$W_t(x,\beta) = \int\limits_{\mathcal{H}_-}^{\sim} e^{\frac{i}{2}\int\limits_{-\infty}^{0} |\frac{d\gamma}{d\tau}|^2 d\tau} \cdot e^{-i\int\limits_{-t}^{0} V(\gamma(\tau)+\beta\tau+x)d\tau} d\gamma \tag{3.36}$$

we get from (3.30) that

$$(e^{-itH}e^{itH_0}\varphi)(x) = \int W_t(x,\beta)e^{i\beta x}d\nu(\beta) , \tag{3.37}$$

where

$$\varphi(x) = \int e^{ix\beta} d\nu(\beta) . \qquad (3.38)$$

Since φ is in $L_2(\mathbb{R}^n)$, which implies that $d\nu$ belongs to $L_1 \cap L_2$, we have, by the assumptions on the potential V, that (3.37) converges strongly. Hence, as $t \to \infty, W_t(x,\beta)$ converges in the strong topology of operators on $L_2(\mathbb{R}^n)$ to a limit $W_+(x,\beta)$, which by (3.37) satisfies

$$(W_+\varphi)(x) = \int W_+(x,\beta) e^{i\beta x} d\nu(\beta) . \qquad (3.39)$$

By the physical interpretation of the wave operators, $W_+(x,\beta)$ as a function of x is the wave function at time zero of the quantum mechanical particle with asymptotic momentum β as $t \to -\infty$. If $m \neq 1$ and $\hbar \neq 1$ one finds easily, by following the previous calculation, that

$$W_t(x,\beta) \int_{\mathcal{H}_-}^{\sim} e^{\frac{im}{2\hbar} \int_{-\infty}^0 |\frac{d\gamma}{d\tau}|^2 d\tau} \cdot e^{-\frac{i}{\hbar} \int_{-t}^0 V(\gamma(\tau)+\frac{\beta}{m}\tau+x)d\tau} d\gamma , \qquad (3.40)$$

and we recall that a particle of momentum β has the classical velocity $\frac{\beta}{m}$. Hence we get the formula for the $W_+(x,\beta)$ of (3.39):

$$W_+(x,mv) = \lim_{t\to\infty} \int_{\mathcal{H}_-}^{\sim} e^{\frac{im}{2\hbar} \int_{-\infty}^0 |\frac{d\gamma}{d\tau}|^2 d\tau} \cdot e^{-\frac{i}{\hbar} \int_{-t}^0 V(\gamma(\tau)+v\tau+x)d\tau} d\gamma . \qquad (3.41)$$

We shall also write this formula as an improper normalized integral

$$W_+(x,mv) = \int_{\mathcal{H}_-}^{\sim} e^{\frac{im}{2\hbar} \int_{-\infty}^0 |\frac{d\gamma}{d\tau}|^2 d\tau} \cdot e^{-\frac{i}{\hbar} \int_{-\infty}^0 V(\gamma(\tau)+v\tau+x)d\tau} d\gamma , \qquad (3.42)$$

keeping in mind that $e^{-\frac{i}{\hbar} \int_{-\infty}^0 V(\gamma(\tau)+v\tau+x)d\tau}$ is not necessarily Fresnel integrable on $\mathcal{H}_-$, and that the integral in (3.42) is defined by the limit (3.41).

On the other hand (3.42) can also be defined without the limit procedure (3.41) in the case where $V(x)$ is a potential such that the perturbation $H = H_0 + V$ is gentle. For gentle perturbations see for instance the references [50] and [51]. One has for example that if $V \in L_1(\mathbb{R}^n) \cap L_\infty(\mathbb{R}^n)$ then the perturbation $H = H_0 + V$ is gentle and, for the case of $\mathbb{R}^3$, one has the stronger result that if $V \in L_{3/2}(\mathbb{R}^3)$ then the perturbation is gentle [[50] Theorems 4, 5 and 6].

In the case when V is gentle and with small gentleness norm, i.e. if for instance $||V||_1$ and $||V||_\infty$ are bounded by a certain constant or, in the case of $\mathbb{R}^3$, if $||V||_{3/2}$ is bounded by a certain constant, then (3.42) may also be defined as follows, setting now again $m = \hbar = 1$:

$$\int\limits_{\mathcal{H}_-}^{\sim} e^{\frac{1}{2}\int\limits_{-\infty}^0 |\frac{d\gamma}{d\tau}|^2 d\tau} \, e^{-i\int\limits_{-\infty}^0 V(\gamma(\tau)+v\tau+x)} \, d\gamma$$

$$= \sum_{n=0}^\infty \frac{(-i)^n}{n!} \int\limits_{\mathcal{H}_-}^{\sim} e^{\frac{1}{2}\int\limits_{-\infty}^0 |\frac{d\gamma}{d\tau}|^2 d\tau} \left(\int\limits_{-\infty}^0 V(\gamma(\tau) + v\tau + x)d\tau \right)^n d\gamma$$

$$= \sum_{n=0}^\infty \frac{(-i)^n}{n!} \int\limits_{-\infty}^0 \cdots \int\limits_{-\infty}^0 \int\limits_{\mathcal{H}_-}^{\sim} e^{\frac{1}{2}\int\limits_{-\infty}^0 |\frac{d\gamma}{d\tau}|^2 d\tau}$$

$$\cdot \quad V(\gamma(t_1) + vt_1 + x)\ldots V(\gamma(t_n) + vt_n + x)d\gamma dt_1 \ldots dt_n \ . \quad (3.43)$$

We have namely that $V(\gamma(t)+vt+x) \in \mathcal{F}(\mathcal{H}_-)$, hence the normalized integral in the last line in (3.43) is well defined. In the same manner as earlier we may compute this normalized integral, and we get by the earlier computations the right hand side of (3.35), with $t = \infty$. Again by the earlier computations we see that this is the same series as the perturbation expansion of W_+ in powers of V, namely

$$\sum_{n=0}^\infty (-i)^n \int\limits_{0 \leq t_1 \leq \ldots \leq t_n} \cdots \int V(t_1) \ldots V(t_n)dt_1 \ldots dt_n \ . \quad (3.44)$$

By the assumption that the perturbation is small and gentle we have that the integrals and the sum in (3.44) actually converge and the sum in (3.44) is equal to the wave operator W_+. Hence the integrals and the sum in (3.43) also converge and the sum is equal to $W_+(x, \beta)$ with $\beta = mv$.

Let us now take $t = 0$ and $s < 0$ in (3.24), and substitute (3.27) for (3.25). We then have

$$\left(e^{-isH_0}e^{isH}\varphi\right)(x)$$

$$= \sum_{n=0}^\infty (-i)^n \int\limits_{s<t_1\leq\ldots\leq t_n\leq 0} \cdots \int \int \cdots \int e^{-\frac{1}{2}\left[\sum\limits_{j,k=1}^n -t_j \vee t_k \alpha_j \alpha_k + 2\delta \sum\limits_{j=1}^n t_j \alpha_j \right]}$$

$$\cdot \quad e^{-i\delta x}d\nu(\beta) \prod_{j=1}^n (d\mu(\alpha_i)dt_i) \ . \quad (3.45)$$

Due to the symmetry of the integrand we get, after a substitution $t_i \to -t_i, i = 1, \ldots, n$:

$$\left(e^{-isH_0}e^{isH}\varphi\right)(x)$$

$$= \sum_{n=0}^{\infty} \frac{(-i)^n}{n!} \int_0^{-s} \cdots \int_0^{-s} \int \cdots \int e^{-\frac{i}{2}\left[\sum_{j,k=1}^{n} t_j \wedge t_k \alpha_j \alpha_k - 2\delta_n \sum_{j=1}^{n} t_j \alpha_j\right]}$$

$$\cdot \ e^{-i\delta_n x} d\nu(\beta) \prod_{i=1}^{n} d\mu(\alpha_i) dt_i \ , \tag{3.46}$$

with $\delta_n = \sum_{i=1}^{n} \alpha_i + \beta$.

For $s < 0$ we now define $W_s^*(\delta, y)$ by

$$\int W_s^*(\delta, y)e^{-i\delta y}\varphi(y)dy = \int (e^{-isH_0}e^{isH}\varphi)(x)e^{-i\delta x}dx \ . \tag{3.47}$$

So that, with ψ in $L_2(\mathbb{R}^n)$ and

$$\psi(x) = \int e^{i\delta x}d\sigma(\delta)$$

$$= \int e^{i\delta x}\hat{\psi}(\delta)d\delta \ , \tag{3.48}$$

we have

$$\iint W_s^*(\delta, y)e^{-i\delta y}\varphi(y)dyd\bar{\sigma}(\delta) = (\psi, e^{-isH_0}e^{isH}\varphi) \ . \tag{3.49}$$

From (3.46) we get

$$\int W_s^*(\delta, y)e^{-i\delta y}\varphi(y)dyd\bar{\sigma}(\delta) = \sum_{m=0}^{\infty} \frac{(-i)^m}{m!} \int_0^{-s} \cdots \int_0^{-s} \int \cdots \int$$

$$\cdot \ e^{-\frac{i}{2}\left[\sum_{j,k=1}^{m} t_j \wedge t_k \alpha_j \alpha_k - 2(\beta + \sum_{j=1}^{m} \alpha_j)\sum_{j=1}^{m} t_j \alpha_j\right]}$$

$$\cdot \ e^{i(\beta + \sum_{j=1}^{m} \alpha_j)x} \bar{\psi}(x)dxd\nu(\beta) \prod_{i=1}^{m} d\mu(\alpha_i)dt_i$$

$$= \sum_{m=0}^{\infty} \frac{(-i)^m}{m!} \int_0^{-s} \cdots \int_0^{-s} \int \cdots \int e^{-\frac{i}{2}\left[\sum_{j,k=1}^{m} t_j \wedge t_k \alpha_j \alpha_k - 2(\beta + \sum_{j=1}^{m} \alpha_j)\sum_{j=1}^{m} t_j \alpha_j\right]}$$

$$\cdot \ (2\pi)^n \bar{\hat{\psi}}(\beta + \sum_{j=1}^{m} \alpha_j)d\nu(\beta) \prod_{i=1}^{m} d\mu(\alpha_i)dt_i \ . \tag{3.50}$$

With the notation $\hat{\varphi}(\beta)d\beta = d\nu(\beta)$ and the substitution $\beta + \sum_{j=1}^{m} \alpha_j \to \delta$ we have that (3.50) is equal to

$$\sum_{m=0}^{\infty} \frac{(-\mathrm{i})^m}{m!} \int_0^{-s} \cdots \int_0^{-s} \int \int \cdots \int e^{-\frac{\mathrm{i}}{2}\left[\sum_{j,k=1}^{m} t_j \wedge t_k \alpha_j \alpha_k - 2\delta \sum_{j=1}^{m} t_j \alpha_j\right]}$$

$$(2\pi)^n \tilde{\psi}(\delta)\hat{\varphi}\left(\delta - \sum_{j=1}^{m} \alpha_j\right) \mathrm{d}\delta \prod_{i=1}^{m} \mathrm{d}\mu(\alpha_i)\mathrm{d}t_i \ .$$

Using now the inverse Fourier transform

$$(2\pi)^n \hat{\varphi}(\delta) = \int \varphi(x)e^{-\mathrm{i}\delta x}\mathrm{d}x \ ,$$

we get (3.50) equal to

$$\sum_{m=0}^{\infty} \frac{(-\mathrm{i})^m}{m!} \int_0^{-s} \cdots \int_0^{-s} \int \int \cdots \int e^{-\frac{\mathrm{i}}{2}\left[\sum_{j,k=1}^{m} t_j \wedge t_k \alpha_j \alpha_k - 2\delta \sum_{j=1}^{m} t_j \alpha_j\right]}$$

$$\cdot\ e^{\mathrm{i}x \sum_{j=1}^{m} \alpha_j} e^{-\mathrm{i}\delta x}\tilde{\psi}(\delta)\varphi(x)\mathrm{d}x\mathrm{d}\delta \prod_{j=1}^{m} \mathrm{d}\mu(\alpha_j)\mathrm{d}t_j \ .$$

Hence we have proved the formula

$$W_s^*(\delta, x) = \sum_{n=0}^{\infty} \frac{(-\mathrm{i})^n}{n!} \int_0^{-s} \cdots \int_0^{-s} \int \int \cdots \int e^{-\frac{\mathrm{i}}{2}\left[\sum_{j,k=1}^{n} t_j \wedge t_k \alpha_j \alpha_k - 2\delta \sum_{j=1}^{n} t_j \alpha_j\right]}$$

$$\cdot\ e^{\mathrm{i}x\left(\sum_{j=1}^{n} \alpha_j\right)} \prod_{j=1}^{n} \mathrm{d}\mu(\alpha_j)\mathrm{d}t_j \ . \tag{3.51}$$

We introduce now the real separable Hilbert space $\mathcal{H}_+$ of continuous functions γ from $[0, \infty]$ to $\mathbb{R}^n$ such that $\gamma(0) = 0$ and $\frac{\mathrm{d}\gamma}{\mathrm{d}\tau}$ is in $L_2([0, \infty], \mathbb{R}^n)$ with norm given by

$$|\gamma|^2 = \int_0^{\infty} \left|\frac{\mathrm{d}\gamma}{\mathrm{d}\tau}\right|^2 \mathrm{d}\tau \ .$$

We verify easily that the function $s \wedge t$ plays the same role in $\mathcal{H}_+$ as the function $-s \vee t$ in $\mathcal{H}_-$, and thus by the same calculations as for $\mathcal{H}_-$ we get that $\int_0^{-s} V(\gamma(\tau) + \delta\tau + x)\mathrm{d}\tau$ is in $\mathcal{F}(\mathcal{H}_+)$ and that

$$W_s^*(\delta, x) = \int_{\mathcal{H}_+} e^{\frac{\mathrm{i}}{2}\int_0^{\infty} \left|\frac{\mathrm{d}\gamma}{\mathrm{d}\tau}\right|^2 \mathrm{d}\tau} e^{-\mathrm{i}\int_0^{-s} V(\gamma(\tau)+\delta\tau+x)\mathrm{d}\tau} \mathrm{d}\gamma \tag{3.52}$$

If m or $\hbar$ are different from 1 we shall define the norm in $\mathcal{H}_+$ by

$$|\gamma|^2 = \frac{m}{\hbar} \int \left|\frac{d\gamma}{d\tau}\right|^2 d\tau$$

and we get the corresponding formula

$$W_s^*(mv, x) = \int_{\mathcal{H}_+} e^{\frac{im}{2\hbar} \int_0^\infty \left|\frac{d\gamma}{d\tau}\right|^2 d\tau} \cdot e^{-\frac{i}{\hbar} \int_0^s V(\gamma(\tau)+v\tau+x)d\tau} d\gamma . \tag{3.53}$$

Let us now assume that the potential is so that the wave operators $W_\pm$ defined by (3.23) exist, then we get from (3.49) that

$$\int W_s^*(\delta, x)e^{-i\frac{\delta}{\hbar}x} \hat{\psi}(\delta)d\delta = \left(e^{-isH} e^{isH_0} \psi\right)(x) \tag{3.54}$$

and the limit of (3.54) as $s \to -\infty$ exists in the strong L_2-sense and defines $W_-^*(\delta, x)$ by

$$\int W_-^*(\delta, x)e^{-i\frac{\delta}{\hbar}x} \hat{\psi}(\delta)d\delta = (W_- \psi)(x) . \tag{3.55}$$

By the physical interpretation of the wave operator W_- we have from (3.55) that $W_-^*(\delta, x)$, as a function of δ for fixed x, is the asymptotic probability amplitude in momentum space as $t \to +\infty$ of a particle located at x for $t = 0$. In the same way as for $W_+(x, \beta)$ we may introduce the improper normalized integral as the limit as $s \to -\infty$ in the weak sense in $\delta = mv$ and x of the normalized integral (3.53) and then write

$$W_-^*(mv, x) = \int_{\mathcal{H}_+} e^{\frac{im}{2\hbar} \int_0^\infty \left|\frac{d\gamma}{d\tau}\right|^2 dt} e^{-\frac{i}{\hbar} \int_0^\infty V(\gamma(\tau)+v\tau+x)d\tau} d\gamma \tag{3.56}$$

or

$$W_-^*(mv, x) = \int_{\gamma(0)=x} e^{\frac{im}{2\hbar} \int_0^\infty \left|\frac{d\gamma}{d\tau}\right|^2 dt} e^{-\frac{i}{\hbar} \int_0^\infty V(\gamma(\tau)+v\tau)d\tau} d\gamma . \tag{3.57}$$

Of course in the case of gentle perturbations (3.56) or (3.57) may also be defined by series expansion of the second term in the integral. Since the scattering operator S is defined by

$$S = W_-^* W_+ \tag{3.58}$$

we get that the scattering amplitude $S(\delta, \beta)$, which is simply the kernel of S in the momentum or Fourier transformed space, is given by the formula

$$S(\delta, \beta) \int_{\mathbb{R}^n} W_-^*(\delta, x) e^{\frac{i}{\hbar}(\beta-\delta)x} W_+(x, \beta) dx \; , \tag{3.59}$$

where we have taken $\hbar \neq 1$ and the integration is to be understood in the weak sense. (3.59) now gives a very interesting and surprising formula for the scattering amplitude

$$S(mv_+, mv_-) = \int_{\lim_{t \to \pm\infty} \frac{\gamma(t)}{t} = V_\pm} e^{\frac{i}{\hbar}(S(\gamma) - S_0(\gamma_0))} d\gamma \tag{3.60}$$

where $\gamma_0(\tau) = v_- \cdot \tau \wedge 0 + v_+ \cdot \tau \vee 0 + y$ are the asymptotes of $\gamma(\tau), S(\gamma)$ is the action along the path γ and $S_0(\gamma_0)$ is the free action along the asymptotic path γ_0. (3.60) then expresses the quantum mechanical scattering amplitude as a normalized integral, over all paths with given asymptotic behavior of $\exp\{\frac{i}{\hbar}(S(\gamma) - S_0(\gamma_0))\}$, where $S(\gamma) - S_0(\gamma_0)$ is the difference of the action along γ and the free action along its asymptotes γ_0. More precisely

$$S(\gamma) - S_0(\gamma_0) = \int_{-\infty}^{\infty} \left(\frac{m}{2} \left| \frac{d\gamma}{d\tau} \right|^2 - V(\gamma) - \frac{m}{2} \left| \frac{d\gamma_0}{d\tau} \right|^2 \right) d\tau \tag{3.61}$$

and with $\gamma = \tilde\gamma + \gamma_0$ we get

$$S(\gamma) - S_0(\gamma_0)$$
$$= \frac{m}{2} \int_{-\infty}^{\infty} \left| \frac{d\tilde\gamma}{d\tau} \right|^2 d\tau - \int_{-\infty}^{\infty} V(\tilde\gamma + \gamma_0) d\tau + mv_+ \int_{0}^{\infty} \frac{d\tilde\gamma}{d\tau} d\tau + mv_- \int_{-\infty}^{0} \frac{d\tilde\gamma}{d\tau} d\tau$$

i.e.

$$S(\gamma) - S_0(\gamma_0) = \frac{m}{2} \int_{-\infty}^{\infty} \left| \frac{d\tilde\gamma}{d\tau} \right|^2 d\tau - \int_{-\infty}^{\infty} V(\tilde\gamma + \gamma_0) d\tau - \tilde\gamma(0)(mv_+ - mv_-). \tag{3.62}$$

Remark. Although $S(\gamma)$ and $S_0(\gamma_0)$ diverge, we see that $S(\gamma) - S_0(\gamma_0)$ is well defined whenever $\tilde\gamma$ is absolutely continuous with derivative in $L_2(R)$, if v_+ and v_- are different from zero and the potential V tends to zero faster than $|x|^{-1-\varepsilon}$ for some positive ε. It is interesting to note that this are just the conditions that are needed for the existence of the scattering amplitude in quantum mechanics.

Let now $\tilde\gamma(0) = x$, we may then set

$$\tilde\gamma(\tau) = \begin{cases} \gamma_+(\tau) + x & \text{for} \quad \tau \geq 0 \\ \gamma_-(\tau) + x & \text{for} \quad \tau \leq 0 \end{cases},$$

where $\gamma_\pm \in \mathcal{H}_\pm$. With this notation we get from (3.62) that

$$S(\gamma) - S_0(\gamma_0) = \frac{m}{2} \int_0^\infty \left|\frac{\mathrm{d}\gamma_+}{\mathrm{d}\tau}\right|^2 \mathrm{d}\tau - \int_0^\infty V(\gamma_+ + v_+\tau + x)\mathrm{d}\tau - x(mv_+ - mv_-)$$

$$+ \frac{m}{2} \int_{-\infty}^0 \left|\frac{\mathrm{d}\gamma_-}{\mathrm{d}\tau}\right|^2 \mathrm{d}\tau - \int_{-\infty}^0 V(\gamma_- + v_\tau + x)\mathrm{d}\tau . \tag{3.63}$$

Hence, using the identity (3.63) we give a precise meaning to the normalized integral (3.60) by the following definition

$$\int\limits_{\lim\limits_{t\to\pm\infty} \frac{\gamma(t)}{t}=V_\pm} \mathrm{e}^{\frac{\mathrm{i}}{\hbar}(S(\gamma)-S_0(\gamma_0))} \mathrm{d}\gamma$$

$$= \int_{\mathbb{R}^n} \mathrm{d}x\, \mathrm{e}^{\frac{\mathrm{i}}{\hbar}mx(v_+-v_-)} \int\limits_{\mathcal{H}_-\oplus\mathcal{H}_+} \mathrm{e}^{\frac{\mathrm{i}m}{2\hbar}\int_{-\infty}^0 \left|\frac{\mathrm{d}\gamma_-}{\mathrm{d}\tau}\right|^2 \mathrm{d}\tau + \frac{\mathrm{i}m}{2\hbar}\int_0^\infty \left|\frac{\mathrm{d}\gamma}{\mathrm{d}\tau}\right|^2 \mathrm{d}\tau}$$

$$\mathrm{e}^{\frac{-\mathrm{i}}{\hbar}\int_{-\infty}^0 V(\gamma_-+v_-\tau+x)\mathrm{d}\tau} \cdot \mathrm{e}^{-\frac{\mathrm{i}}{\hbar}\int_0^\infty V(\gamma_++v_+\tau+x)\mathrm{d}\tau} \mathrm{d}\gamma , \tag{3.64}$$

where the normalized integral above is the normalized integral on $\mathcal{H} = \mathcal{H}_+ \oplus \mathcal{H}_-$ and $\gamma = \gamma_+ \oplus \gamma_-$.

From (3.64), (3.59), (3.56) and (3.42) we have proved (3.60) as defined by (3.64). We formulate now these results in the following theorem.[5]

Theorem 3.2. *Let the potential V be the Fourier transform of a bounded complex measure and also in a class of potentials such that the wave operators (3.23) exist. Then the wave amplitudes $W_+(x, \beta)$ and $W_-^*(\delta, x)$ defined by (3.39) and (3.55) are given by the following improper normalized integrals*

$$W_+(x, mv) = \int_{\mathcal{H}_-} \mathrm{e}^{\frac{\mathrm{i}m}{2\hbar}\int_{-\infty}^0 \left|\frac{\mathrm{d}\gamma}{\mathrm{d}\tau}\right|^2 \mathrm{d}\tau} \mathrm{e}^{-\frac{\mathrm{i}}{\hbar}\int_{-\infty}^0 V(\gamma(\tau)+v\tau+x)\mathrm{d}\tau} \mathrm{d}\gamma$$

and

$$W_-^*(mv, x) = \int_{\mathcal{H}_+} \mathrm{e}^{\frac{\mathrm{i}m}{2\hbar}\int_0^\infty \left|\frac{\mathrm{d}\gamma}{\mathrm{d}\tau}\right|^2 \mathrm{d}\tau} \mathrm{e}^{-\frac{\mathrm{i}}{\hbar}\int_0^\infty V(\gamma(\tau)+v\tau+x)\mathrm{d}\tau} \mathrm{d}\gamma ,$$

where the improper normalized integrals are defined as the limits of the corresponding ordinary normalized integrals with the integrals over the half lines

---

[5] With obvious modifications the results hold also for the N particle system of Footnote 2 of this chapter.

of the function $V(\gamma(\tau) + v\tau + x)$ substituted by the integrals over finite time intervals of the same function. $\mathcal{H}_+$ is the real separable Hilbert space of absolutely continuous functions γ from $[0, \infty]$ to $\mathbb{R}^n$ such that $\gamma(0) = 0$ and $\frac{d\gamma}{d\tau}$ is in $L_2(R)$, with norm $|\gamma|_+^2 = \frac{m}{\hbar} \int_0^\infty \left|\frac{d\gamma}{d\tau}\right|^2 d\tau$. $\mathcal{H}_-$ is the corresponding space of functions on $[-\infty, 0]$. Moreover the scattering amplitude $S(\alpha, \beta)$, defined as the Fourier transform of the kernel of the scattering operator S, is given by the formula

$$S(mv_+, mv_-) = \int\limits_{\substack{\lim\limits_{t \to \pm\infty} \frac{\gamma(t)}{t} = v_\pm}}^{\sim} e^{\frac{i}{\hbar}(S(\gamma) - S_0(\gamma_0))} d\gamma ,$$

where

$$S(\gamma) - S_0(\gamma_0) = \int\limits_{-\infty}^{\infty} \left(\frac{m}{2} \left|\frac{d\gamma}{d\tau}\right|^2 - V(\gamma) - \frac{m}{2} \left|\frac{d\gamma_0}{d\tau}\right|^2 \right) d\gamma$$

with $\gamma_0(\tau) = v_- \cdot \tau \wedge 0 + v_+ \cdot \tau \vee 0 + y$, and the normalized integral above is defined by (3.64).

Notes

The approach in this section was introduced in the first edition of this book. It was used in an essential way for the development of the method of stationary phase in infinite dimensions and the study of the semiclassical limit (for solutions of the Schrödinger equation in finite time) in [87] and subsequent papers like [69, 70]. Another development, in the case of Hamiltonians with discrete spectrum, is connected with the trace formula, see [66, 67]. The approach concerning scattering theory has not been exploited much further, it gave however inspiration for further work, see, e.g. [186, 187].

4

The Fresnel Integral Relative
to a Non-singular Quadratic Form

In Theorem 3.1 we obtained the solution $\psi(x,t)$ of the Schrödinger equation with initial values $\varphi(x)$ and potential $V(x)$ in the form

$$\psi(x,t) = \overset{\sim}{\underset{\gamma(t)=x}{\int}} e^{\frac{im}{2\hbar}\int_0^t |\frac{d\gamma}{d\tau}|^2 d\tau} e^{-\frac{i}{\hbar}\int_0^t V(\gamma(\tau))d\tau} \varphi(\gamma(0))d\gamma \qquad (4.1)$$

$$\overset{\text{Def}}{=} \overset{\sim}{\underset{\mathcal{H}}{\int}} e^{\frac{im}{2\hbar}\int_0^t |\frac{d\gamma}{d\tau}|^2 d\tau} e^{-\frac{i}{\hbar}\int_0^t V(\gamma(\tau)+x)d\tau} \varphi(\gamma(0)+x)d\gamma \;,$$

where $\mathcal{H}$ was the real Hilbert space of continuous paths γ such that $\gamma(t) = 0$ and with norm square given by $\frac{m}{\hbar}\int_0^t |\frac{d\gamma}{d\tau}|^2 d\tau$, if both V and φ are Fourier transforms of bounded complex measures. If we are, and we shall be, interested in the anharmonic oscillator, then we must deal with potentials of the form

$$V'(x) = \frac{1}{2}xA^2 x + V(x) \;, \qquad (4.2)$$

where $xA^2 x$ is a strictly positive definite form on $\mathbb{R}^n$, corresponding to the strictly positive definite symmetric linear transformation A^2 on $\mathbb{R}^n$, and $V(x)$ is a nice function, which we shall take to be in the class of Fourier transforms of bounded complex measures. For such potentials, of the form (4.2), we can not prove (4.1) in the same way as in the previous section, because if we substitute V' for V in (4.1) we do not get a Fresnel integrable function on $\mathcal{H}$ and the formula therefore does not make sense as it stands. On the other hand, we may write (4.1) with V' instead of V in the following manner

$$\overset{\sim}{\underset{\gamma(t)=x}{\int}} e^{\frac{im}{2\hbar}\int_0^t \dot{\gamma}(\tau)^2 d\tau - \frac{i}{2\hbar}\int_0^t \gamma(\tau)A^2\gamma(\tau)d\tau} e^{-\frac{i}{\hbar}\int_0^t V(\gamma(\tau))d\tau} \varphi(\gamma(0))d\gamma \;, \qquad (4.3)$$

where $\frac{d\gamma}{d\tau}(\tau) = \dot{\gamma}(\tau)$.

For small values of t

$$\frac{m}{\hbar} \int\limits_0^t \dot\gamma(\tau)^2 \mathrm{d}\tau - \frac{1}{\hbar} \int\limits_0^t \gamma(\tau) A^2 \gamma(\tau) \mathrm{d}\tau \tag{4.4}$$

is namely a strictly positive definite quadratic form on the space of continuous paths such that $\gamma(t) = 0$. Hence we may introduce the real Hilbert space $\mathcal{H}'$ of continuous functions from $[0, t]$ to $\mathbb{R}^n$ with $\gamma(t) = 0$ such that (4.4) is bounded with (4.4) as the norm square, and define (4.3) by

$$\widetilde{\int\limits_{\mathcal{H}'}} \mathrm{e}^{\frac{i}{2}\left[\frac{m}{\hbar}\int\limits_0^t \dot\gamma(\tau)^2 \mathrm{d}\tau - \frac{1}{\hbar}\int\limits_0^t (\gamma(\tau)+x)A^2(\gamma(\tau)+x)\mathrm{d}\tau\right]} \mathrm{e}^{-\frac{i}{\hbar}\int\limits_0^t V(\gamma(\tau)+x)\mathrm{d}\tau} \varphi(\gamma(0)+x)\mathrm{d}\gamma \,, \tag{4.5}$$

where the normalized integral $\widetilde{\int}_{\mathcal{H}'}$ is the one defined in Chap. 2.

One verifies then easily that

$$\mathrm{e}^{-\frac{i}{\hbar}xA^2\int\limits_0^t \gamma(\tau)\mathrm{d}\tau} \, \mathrm{e}^{-\frac{i}{\hbar}\int\limits_0^t V(\gamma(\tau)+x)\mathrm{d}\tau} \varphi(\gamma(0)+x) \cdot \mathrm{e}^{-\frac{it}{2\hbar}xA^2x} \tag{4.6}$$

is Fresnel integrable on $\mathcal{H}'$ and that, up to a constant, the Fresnel integral (4.5) gives the solution of the corresponding Schrödinger equation at time t. The constant, which depends only on t, m and A^2, comes from the fact that the normalized integral is defined by a normalization given by the inner product in the Hilbert space. In this way we can thus prove an analogue of the Feynman–Ito formula also for the anharmonic oscillator. But this formula would then only hold for small values of t. This is of course rather unsatisfactory, and we shall therefore not give the detailed proof here.

If on the other hand we want to make sense out of (4.3) not only for small t, we must define the Fresnel integral also for the case where the quadratic form in question is not necessarily positive definite any longer. We shall see that this is possible and in this way make sense of (4.3) not only for small values of t. To define this extension of the Fresnel integral we shall introduce a densely defined symmetric operator B in the separable real Hilbert space $\mathcal{H}$ of Chap. 2, and the quadratic form is then given by

$$(x, Bx) \tag{4.7}$$

for $x \in D(B) \subset \mathcal{H}$. It is also necessary to assume that B is non degenerate in some suitable sense, and we shall here assume that B is non degenerate in the following sense. There exists a dense subspace D of $\mathcal{H}$ such that D contains the range of B and there exists a symmetric bilinear form $\triangle(x, y)$ defined on $D \times D$ such that Im $\triangle(x, x) \leq 0$ and

$$\triangle(x, By) = (x, y) \tag{4.8}$$

for all $x \in D$ and all y in the domain $D(B)$ of B. $\triangle$ is in the above sense an inverse form of the form (4.7), and in the case where $\mathcal{H}$ is finite dimensional the existence of $\triangle$ is equivalent to B having an inverse and $\triangle$ is in that case given by the inverse matrix of B. We shall further assume that D is a separable Banach space with norm $||x||$ which is stronger than the norm $|x|$ in $\mathcal{H}$, i.e.

$$|x| \leq a||x|| \tag{4.9}$$

for all $x \in D$, and we shall assume that the form $\triangle(x, y)$ is a continuous symmetric bilinear form on D. From (4.9) and the self duality of $\mathcal{H}$ we get the natural embedding

$$D \subset \mathcal{H} \subset D^* , \tag{4.10}$$

where D^* is the dual space of D. The embedding $\mathcal{H} \subset D^*$ is just the restriction mapping i.e. by restricting a continuous linear function on $\mathcal{H}$ to D we get, by (4.9), a continuous linear mapping on D. It also follows from (4.9) that $||x||* \leq a|x|$, where $||x||*$ is the norm in D^*. Hence all the injections in (4.10) are continuous. In the general case $\triangle$ is not uniquely given by B. However if B is non degenerate in the stronger sense that it has a bounded continuous inverse B^{-1} on $\mathcal{H}$, then it follows easily that $D = \mathcal{H}$, since D contains the range of B, and that $\triangle(x, y) = (x, B^{-1}y)$. Hence $\triangle$ is in this case unique and real.

Since $\triangle(x, y)$ is continuous on $D \times D$ we have that, for fixed x, $\triangle(x, y)$ is a continuous complex linear functional on D, hence in D^*. This gives then a mapping of D into D^* which is, by (4.8), a left inverse of B, considered as a map from $D(B) \subset D^*$ into D.

We now define the Fresnel integral with respect to the form $\triangle$.

Definition 4.1. *Let D be a real separable Banach space with norm $|| \ ||$ and let $\mathcal{H}$ be a real separable Hilbert space with inner product $(,)$ and norm $| \ |$, such that D is densely contained in $\mathcal{H}$ and the norm in D is stronger than the norm in $\mathcal{H}$. Let B be a densely defined symmetric operator on $\mathcal{H}$ such that the range of B is contained in D and let $\triangle(x, y)$ be a symmetric and continuous bilinear form on $D \times D$ such that Im $\triangle(x, x) \leq 0$, and $\triangle(x, By) = (x, y)$ for all $x \in D$ and $y \in D(B)$. The space $\mathcal{F}(D^*)$ of Fresnel integrable functions on D^* is the space of Fourier transforms of bounded complex measures on D. For any $f \in \mathcal{F}(D^*)$ we get, by the inclusion $\mathcal{H} \subset D^*$, that*

$$f(x) = \int_D e^{i(x,y)} d\mu(y) \tag{4.11}$$

for $x \in \mathcal{H}$. We now define, for any $f \in \mathcal{F}(D^*)$, the Fresnel integral with respect to $\triangle$ by

$$\int_{\mathcal{H}}^{\triangle} e^{\frac{1}{2}(x,Bx)} f(x) dx = \int_D e^{-\frac{1}{2}\triangle(x,x)} d\mu(x) . \tag{4.12}$$

The integral on the right hand side is well defined since Im $\triangle(x, x) \le 0$ and $\triangle(x, x)$ is continuous on D. We shall also call $\int_{\mathcal{H}}^{\triangle} e^{\frac{i}{2}(x,Bx)} f(x)\mathrm{d}x$ the integral normalized with respect to $\triangle$ of the function $e^{\frac{i}{2}(x,Bx)} f(x)$, and also use the notation $\mathcal{F}_{\triangle}(f)$ for this integral.

We have now the following proposition.

Proposition 4.1. *The space of Fresnel integrable functions $\mathcal{F}(D^*)$ is a Banach-function-algebra in the norm $||f||_0 = ||\mu||$ and $\mathcal{F}(D^*) \subseteq \mathcal{F}(\mathcal{H})$. $\mathcal{F}_{\triangle}(f)$ is a bounded continuous linear functional on $\mathcal{F}(D^*)$ such that $|\mathcal{F}_{\triangle}(f)| \le ||f||_0$ and normalized such that $\mathcal{F}_{\triangle}(1) = 1$. The condition $\triangle(x, By) = (x, y)$ for $x \in D$ and $y \in D(B)$ implies that B^{-1} is well defined and $\triangle(x, y) = (x, B^{-1}y)$ for $x \in D$ and y in the range of B. Hence, if the range of B is dense in D, $\triangle(x, y)$ is uniquely given by B and is the continuous extension of the form $(x, B^{-1}y)$. If $B \ge a1$ with $a > 0$ and the range of B dense in D, then $\mathcal{F}(D^*) \subseteq \mathcal{F}(\mathcal{H}_B)$ and*

$$\int_{\mathcal{H}}^{\triangle} e^{\frac{i}{2}(x,Bx)} f(x)\mathrm{d}x = \int_{\mathcal{H}_B}^{\sim} e^{\frac{i}{2}|x|_B^2} f(x)\mathrm{d}x \ ,$$

with $|x|_B^2 = (x, Bx)$ and where $\mathcal{H}_B$ is the closure of $D(B)$ in the norm $| \ |_B$.

Proof. Let $f \in \mathcal{F}(D^*)$, then, for $x \in \mathcal{H}$,

$$f(x) = \int_D e^{\mathrm{i}(x,y)}\mathrm{d}\mu(y) \ .$$

Since the D-norm is stronger than the $\mathcal{H}$-norm, we have that the $\mathcal{H}$-norm $|x|$ is a continuous function on D, from which it follows that any $\mathcal{H}$-continuous function is also D-continuous, so the restriction of the integral with respect to μ from $C(D)$ to $C(\mathcal{H})$ gives a measure on $\mathcal{H}$, which we shall denote by $\mu_{\mathcal{H}}$. Since $e^{\mathrm{i}(x,y)}$ for $x \in \mathcal{H}$ is in $C(\mathcal{H})$, we therefore have that, by the definition of $\mu_{\mathcal{H}}$,

$$f(x) = \int_{\mathcal{H}} e^{\mathrm{i}(x,y)}\mathrm{d}\mu_{\mathcal{H}}(y) \ , \qquad\qquad (4.13)$$

hence that $f \in \mathcal{F}(\mathcal{H})$. That $||\mu_{\mathcal{H}}|| = ||\mu||$ is obvious, so that $\mathcal{F}(D^*)$ is a Banach subspace of $\mathcal{F}(\mathcal{H})$. That it is also a Banach algebra follows as in Proposition 2.2 from the fact that D is a separable metric group. The bound $|\mathcal{F}_{\triangle}(f)| \le ||f||_0$ follows from the fact that $\mathrm{Im}\triangle(x, x) \le 0$, and $\mathcal{F}_{\triangle}(1) = 1$ is obvious from the definition (4.12). From

$$\triangle(x, By) = (x, y) \ , \qquad\qquad (4.14)$$

for $x \in D$ and any $y \in D(B)$, we get that $By = 0$ implies that $y = 0$, since D is dense in $\mathcal{H}$. Hence B^{-1} is well defined with domain equal to the range of B. Hence

$$\triangle(x, y) = (x, B^{-1}y) \tag{4.15}$$

for $x \in D$ and y in the range of B. So that, if the range of B is dense in D, then $\triangle(x, y)$ is uniquely given by (4.15) and therefore also real.

If $B \geq a1$ with $a > 0$ and the range of B is dense in D, then, for $x \in R(B)$, the range of B,

$$|B^{-1}x|_B^2 = (x, B^{-1}x) = \triangle(x, x) , \tag{4.16}$$

which is bounded and continuous in the D-norm. Hence B^{-1} maps $R(B)$ into $\mathcal{H}_B$, continuously in the D-norm on $R(B)$. Since $R(B)$ is dense in D it has a unique continuous extension, which we shall also denote by B^{-1}, such that B^{-1} maps D into $\mathcal{H}_B$ boundedly. To prove the identity in the proposition we first prove that $\mathcal{F}(D^*) \subseteq \mathcal{F}(\mathcal{H}_B)$. Let $f \in \mathcal{F}(D^*)$, then for $x \in \mathcal{H}$ we have

$$f(x) = \int_D e^{i(x,y)} d\mu(y) . \tag{4.17}$$

Let $g \in C(\mathcal{H}_B)$, we define μ_B by

$$\int_{\mathcal{H}_B} g(x) d\mu_B(x) = \int_D g(B^{-1}x) d\mu(x) . \tag{4.18}$$

$g(B^{-1}x) \in C(D)$ since $B^{-1} : D \to \mathcal{H}_B$ continuously. If $x \in D(B)$ then by (4.17) and (4.18) we have

$$f(x) = \int_D e^{i(Bx, B^{-1}y)} d\mu(y)$$

$$= \int_{\mathcal{H}_B} e^{i(Bx, y)} d\mu_B(y) ,$$

so that

$$f(x) = \int_{\mathcal{H}_B} e^{i(x,y)_B} d\mu_B(y) . \tag{4.19}$$

Now (4.19) holds for all $x \in D(B)$ and $D(B)$ is by definition dense in $\mathcal{H}_B$ in the $\mathcal{H}_B$-norm. On the other hand the right hand side of (4.19) is obviously uniformly continuous in the $\mathcal{H}_B$-norm. But, from (4.17), $f(x)$, for $x \in \mathcal{H}$, is uniformly continuous in the $\mathcal{H}$-norm, which is weaker than the $\mathcal{H}_B$-norm. Hence by unique extension of uniformly continuous functions defined on dense subspaces, (4.19) must hold for all x in $\mathcal{H}_B$. This then proves that, in this case, $\mathcal{F}(D^*) \subset \mathcal{F}(\mathcal{H}_B)$, and by (4.19.) we have

$$\int_{\mathcal{H}_B}^{\sim} e^{\frac{i}{2}|x|_B^2} f(x)\mathrm{d}x = \int_{\mathcal{H}_B} e^{-\frac{i}{2}|x|_B^2} \mathrm{d}\mu_B(x)$$

$$= \int_D e^{-\frac{i}{2}|B^{-1}x|_B^2} \mathrm{d}\mu(x) .$$

On the other hand we have (4.16) for all $x \in R(B)$, but by the continuity of $B^{-1} : D \to \mathcal{H}_B$ and the continuity of $\triangle(x, x)$ in the D-norm we have that $|B^{-1}x|_B = \triangle(x, x)$ for all x in D, hence

$$\int_{\mathcal{H}_B}^{\sim} e^{\frac{i}{2}|x|_B^2} f(x)\mathrm{d}x = \int_D e^{-\frac{i}{2}\triangle(x,x)} \mathrm{d}\mu(x) ,$$

which by (4.17) proves the identity in the proposition.   □

Proposition 4.2. *The integral normalized with respect to $\triangle$ is invariant under translations by vectors in the domain of B, i.e., for $y \in D(B)$,*

$$\int_{\mathcal{H}}^{\triangle} e^{\frac{i}{2}(x+y,B(x+y))} f(x + y)\mathrm{d}x = \int_{\mathcal{H}}^{\triangle} e^{\frac{i}{2}(x,Bx)} f(x)\mathrm{d}x .$$

If $\mathcal{H}$ is finite dimensional then B^{-1} is bounded and $\triangle$ is uniquely given by $\triangle(x, y) = (x, B^{-1}y)$ for all x and y, and with $\mathcal{H} = \mathbb{R}^n$

$$\int_{\mathcal{H}}^{\triangle} e^{\frac{i}{2}(x,Bx)} f(x)\mathrm{d}x = |(1/2\pi i)B|^{+\frac{1}{2}} \int_{\mathbb{R}^n} e^{\frac{i}{2}(x,Bx)} f(x)\mathrm{d}x$$

for any $f \in \mathcal{S}(\mathbb{R}^n)$, where $|(1/2\pi i)B|$ is the determinant of the transformation $(1/2\pi i)B$ in $\mathbb{R}^n$, and the integral on the right hand side is the Lebesgue integral.[1]

---

[1] The square root $|(1/2\pi i)B|^{\frac{1}{2}}$ is given by the formula

$$|(1/2\pi i)B|^{\frac{1}{2}} = (1/2\pi)^{n/2}\|B\|^{\frac{1}{2}}e^{-i\frac{\pi}{4}\mathrm{sign}B} ,$$

where $(2\pi)^{n/2}$ is the positive root, $\|B\|^{\frac{1}{2}}$ is the positive root of the absolute value of the determinant of B and sign B is the signature of the form B. This is according to the formula

$$|(1/2\pi i)B|^{\frac{1}{2}} \int_{\mathbb{R}^n} e^{\frac{i}{2}(x,Bx)} \mathrm{d}x = 1 .$$

If $\mathcal{H}$ is infinite dimensional and B^{-1} is bounded and everywhere defined, then $D = \mathcal{H}$, their norms are equivalent and $\triangle$ is uniquely given by $\triangle(x, y) = (x, B^{-1}y)$ for all x and y. Moreover B is self adjoint and $\mathcal{H} = \mathcal{H}_+ \oplus \mathcal{H}_-$, with $B = B_+ \ominus B_-$, where $B_\pm \geq a1$ for some $a > 0$, and

$$\overset{\triangle}{\underset{\mathcal{H}}{\int}} e^{\frac{i}{2}(x,Bx)} f(x)\mathrm{d}x = \underset{\mathcal{H}_{B_+}}{\int} e^{\frac{i}{2}|x_1|^2_{B_+}} \left[\overline{\underset{\mathcal{H}_{B_-}}{\int} e^{\frac{i}{2}|x_2|^2_{B_-}} \overline{f(x_1, x_2)}\mathrm{d}x_2} \right] \mathrm{d}x_1 \, ,$$

where $f(x_1, x_2) = f(x_1 \oplus x_2)$, $\mathcal{H}_{B_+}$ is the closure of $D(B_\pm) \subset \mathcal{H}_+$ in the norm $|x|^2_{B_\pm} = (x, B_\pm x)$, and the integrals on the right hand side are ordinary normalized integrals as defined in Chap. 2.

Proof. Let $y \in D(B)$, then

$$e^{\frac{i}{2}(x+y,B(x+y))} f(x + y) = e^{\frac{i}{2}(y,By)} e^{\frac{i}{2}(x,Bx)} e^{i(x,By)} f(x + y) \, , \qquad (4.20)$$

so we shall first prove that $e^{i(x,By)} f(x + y) \in \mathcal{F}(D^*)$. We have

$$e^{i(x,By)} f(x + y) = e^{i(x,By)} \int_D e^{i(x+y,z)}\mathrm{d}\mu(z)$$

$$= \int_D e^{i(x,z)} e^{i(x,By)} e^{i(y,z)}\mathrm{d}\mu(z)$$

$$= \int_D e^{i(x,z+By)} e^{i(y,z)}\mathrm{d}\mu(z) \, ,$$

and since $By \in D$, so that $z \to z - By$ is a continuous transformation in D, we get

$$e^{i(x,By)} f(x + y) = \int_D e^{i(x,z)} e^{i(y,z-By)}\mathrm{d}\mu(z - By) \, , \qquad (4.21)$$

which is obviously in $\mathcal{F}(D^*)$. Hence by (4.20), (4.21) and the definition of the integral normalized with respect to $\triangle$ we get

$$\overset{\triangle}{\underset{\mathcal{H}}{\int}} e^{\frac{i}{2}(x+y,B(x+y))} f(x + y)\mathrm{d}x = e^{\frac{i}{2}(y,By)} \int_D e^{-\frac{i}{2}\triangle(x,x)} e^{i(y,x-By)}\mathrm{d}\mu(x - By) \, ,$$

which by the definition of the measure $\mathrm{d}\mu(x - By)$ is

$$e^{\frac{i}{2}(y,By)} \int_D e^{-\frac{i}{2}\triangle(x+By,x+By)} e^{i(y,x)}\mathrm{d}\mu(x)$$

$$= e^{\frac{i}{2}(y,By)} e^{-\frac{i}{2}\triangle(By,By)} \int_D e^{-\frac{i}{2}\triangle(x,x)} e^{-i\triangle(x,By)} e^{i(y,x)}\mathrm{d}\mu(x)$$

and, using now that for $y \in D(B)$ and $x \in D$ we have $\triangle(x, By) = (x, y)$, we get

$$\int e^{\frac{i}{2}\triangle(x,x)} d\mu(x) .$$

This proves the translation invariance. We have proved previously that $By = 0 \Rightarrow y = 0$, which in the finite dimensional case implies that B is onto and that B^{-1} is bounded. In this case, since D is dense in $\mathcal{H}$, we have that $D = \mathcal{H}$ and therefore that $\triangle(x, y) = (x, B^{-1}y)$ for all x and y. Let now $f(x) \in \mathcal{S}(\mathbb{R}^n), \mathbb{R}^n = \mathcal{H}$, then

$$\int_{\mathcal{H}}^{\triangle} e^{\frac{i}{2}(x,Bx)} f(x)\mathrm{d}x \overset{\mathrm{Def}}{=} \int e^{-\frac{i}{2}(x,B^{-1}x)} \hat{f}(x)\mathrm{d}x \tag{4.22}$$

with

$$f(x) = \int e^{\mathrm{i}(x,y)} \hat{f}(y)\mathrm{d}y .$$

On the other hand one verifies easily that, for $f \in \mathcal{S}(\mathbb{R}^n)$,

$$\int e^{-\frac{i}{2}(x,B^{-1}x)} \hat{f}(x)\mathrm{d}x = |(1/2\pi\mathrm{i})B|^{+\frac{1}{2}} \int e^{\frac{i}{2}(x,Bx)} f(x)\mathrm{d}x . \tag{4.23}$$

This proves the second part of the proposition. Let now $\mathcal{H}$ be infinite dimensional and B^{-1} bounded and everywhere defined. Then the range of B is $\mathcal{H}$, hence $D = \mathcal{H}$, so that the D-norm $||x||$ is a norm on $\mathcal{H}$ which is everywhere defined, hence by the general theory of functional analysis it is bounded with respect to the norm in $\mathcal{H}$. Therefore the D-norm and $\mathcal{H}$ norm are equivalent. Moreover since B is the inverse of a bounded symmetric operator, it is self adjoint and let $\mathcal{H} = \mathcal{H}_+ \oplus \mathcal{H}_-$ be the spectral decomposition of B in the subspaces where B is positive and B is negative. This decomposition is unique since we already know that zero is not an eigenvalue of B. Let $\pm B_\pm$ be the restrictions of B to $\mathcal{H}_\pm$. By the spectral representation theorem for self adjoint operators it follows from B^{-1} bounded that $B_\pm \geq a1$ for some $a > 0$. Let now $f \in \mathcal{F}(D^*)$, then since $D = \mathcal{H}$ and D and $\mathcal{H}$ are equivalent as metric spaces and the fact that $\mathcal{H}$ and $\mathcal{H}_+ \times \mathcal{H}_-$ are equivalent as metric spaces, we get that any measure on D may be regarded as a measure on $\mathcal{H}_+ \times \mathcal{H}_-$ and therefore, for $x = x_1 \oplus x_2 \in \mathcal{H}$:

$$f(x) = \int_{\mathcal{H}_+ \times \mathcal{H}_-} e^{\mathrm{i}(x_1,y_1)} e^{\mathrm{i}(x_2,y_2)} \mathrm{d}\mu(y_1, y_2) . \tag{4.24}$$

Consider now the measure μ_{x_1} on $\mathcal{H}_-$ defined by

$$\int_{\mathcal{H}_-} g(y_2)\mathrm{d}\mu_{x_1}(y_2) = \int_{\mathcal{H}_+ \times \mathcal{H}_-} g(-y_2)e^{-\mathrm{i}(x_1,y_1)} \overline{\mathrm{d}\mu(y_1, y_2)} , \tag{4.25}$$

for any $g \in C(\mathcal{H}_-)$. Then

$$\overline{f(x_1, x_2)} = \int_{\mathcal{H}_-} e^{i(x_2, y_2)} d\mu_{x_1}(y_x) \ . \tag{4.26}$$

Let now $\mathcal{H}_{B_\pm}$ be the completion of $D(B) \cap \mathcal{H}_\pm$ in the norm $|x|_{B_\pm}^2 = (x, x)_{B_\pm} = (x, B_\pm x)$. Then for $x_2 \in D(B) \cap \mathcal{H}_-$ we have

$$\overline{f(x_1, x_2)} = \int_{\mathcal{H}_-} e^{i(x_2, B^{-1} y_2)_{B_-}} d\mu_{x_1}(y_2)$$

i.e.

$$\overline{f(x_1, x_2)} = \int_{\mathcal{H}_{B_-}} e^{i(x_2, y_2)_{B_-}} d\mu_{x_1}^{B_-}(y_2) \ , \tag{4.27}$$

where $d\mu_{x_1}^{B_-}$ is the measure on $\mathcal{H}_{B_-}$ defined by

$$\int_{\mathcal{H}_{B_-}} g(y) d\mu_{x_1}^{B_-}(y) = \int_{\mathcal{H}_-} g(B_-^{-1} y) d\mu_{x_1}(y) \ , \tag{4.28}$$

for any $g \in C(\mathcal{H}_B)$. (4.28) defines a measure on $\mathcal{H}_{B_-}$, since

$$|B_-^{-1} y|_{B_-}^2 = (y, B_-^{-1} y) \le a||y||^2 \tag{4.29}$$

so that B_-^{-1} maps $\mathcal{H}_-$ into $\mathcal{H}_{B_-}$ continuously. By (4.27), for fixed x_1, $\overline{f(x_1, x_2)} \in \mathcal{F}(\mathcal{H}_{B_-})$. Hence we may compute the inner integral in Proposition 4.2, and we get for fixed x_1

$$g(x_1) = \int_{\mathcal{H}_{B_-}}^{\sim} e^{\frac{i}{2}|x_2|_{B_-}^2} \overline{f(x_1, x_2)} dx_2$$

$$= \int_{\mathcal{H}_{B_-}} e^{-\frac{i}{2}|x_2|_{B_-}^2} d\mu_{x_1}^{B_-}(x_2)$$

$$= \int_{\mathcal{H}_-} e^{-\frac{i}{2}|B_-^{-1} x_2|_{B_-}^2} d\mu_{x_1}(x_2)$$

$$= \int_{\mathcal{H}_-} e^{-\frac{i}{2}(x_2, B_-^{-1} x_2)} d\mu_{x_1}(x_2) \ , \tag{4.30}$$

which by the definition of $d\mu_{x_1}$ is equal to

$$\int_{\mathcal{H}_+\times\mathcal{H}_-} e^{-\frac{1}{2}(y_2,B_-^{-1}y_2)}e^{-i(x_1,y_1)}\overline{d\mu(y_1,y_2)} \; .$$

Hence

$$\overline{g(x_1)} = \int_{\mathcal{H}_+\times\mathcal{H}_-} e^{i(x_1,y_1)}e^{\frac{1}{2}(y_2,B_-^{-1}y_2)}d\mu(y_1,y_2) \; . \tag{4.31}$$

Define now the measure $d\nu(y_1)$ on $\mathcal{H}_+$ by

$$\int_{\mathcal{H}_+} h(y_1)d\nu(y_1) = \int_{\mathcal{H}_+\times\mathcal{H}_-} h(y_1)e^{\frac{1}{2}(y_2,B_-^-y_2)}d\mu(y_1,y_2) \; , \tag{4.32}$$

then

$$\overline{g(x_1)} = \int_{\mathcal{H}_+} e^{i(x_1,y_1)}d\nu(y_1)$$

$$= \int_{\mathcal{H}_+} e^{i(x_1,B_+^{-1}y_1)B_+}d\nu(y_1) \; .$$

Hence

$$\overline{g(x_1)} = \int_{\mathcal{H}_{B_+}} e^{i(x_1,y_1)}d\nu^B(y_1) \; ,$$

so that $\bar{g} \in \mathcal{F}(\mathcal{H}_{B_+})$ and we may therefore compute the outer integral in the proposition and we get

$$\int_{\mathcal{H}_{B_+}} e^{\frac{i}{2}|x_1|_{B_+}^2}\overline{g(x_1)}dx_1 = \int_{\mathcal{H}_{B_+}} e^{-\frac{i}{2}|y_1|_{B_+}^2}d\nu^{B_+}(y_1)$$

$$= \int_{\mathcal{H}_+} e^{-\frac{i}{2}|B_+^{-1}y_1|_{B_+}^2}d\nu(y_1)$$

$$= \int_{\mathcal{H}_+} e^{-\frac{i}{2}(y_1,B_+^{-1}y_1)}d\nu(y_1) \; .$$

By the definition (4.32) of $d\nu(y_1)$ we get this equal to

$$\int_{\mathcal{H}_+\times\mathcal{H}_-} e^{-\frac{i}{2}(y_1,B_+^{-1}y_1)}e^{\frac{i}{2}(y_2,B_-^{-1}y_2)}d\mu(y_1,y_2)$$

$$= \int_{\mathcal{H}} e^{-\frac{i}{2}(y,B^{-1}y)}d\mu(y) = \int_D e^{-\frac{i}{2}\triangle(y,y)}d\mu(y) \; .$$

which by definition is the left hand side of the last equality in the proposition. This then completes the proof of this proposition.   □

The case where B is a bounded symmetric operator with $D(B) = R(B) = \mathcal{H}$, such that B^{-1} is bounded, deserves special attention. We have seen that in this case the space D must be equal to $\mathcal{H}$ and the form $\triangle(x,y)$ is unique and equal to $(x, B^{-1}y)$. Since $\triangle$ is unique we may drop it in the notation of the integral normalized with respect to $\triangle$ and we shall simply write

$$\widetilde{\int_{\mathcal{H}}} e^{\frac{i}{2}\langle x,x\rangle} f(x)\mathrm{d}x \stackrel{\text{Def}}{=} \overset{\triangle}{\int_{\mathcal{H}}} e^{\frac{i}{2}(x,Bx)} f(x)\mathrm{d}x \ , \tag{4.33}$$

where $\langle x,y\rangle = (x, By)$ and $\triangle(x,y) = (x, B^{-1}y)$, in the case B and B^{-1} are both bounded with domains equal to $\mathcal{H}$. In this case we have, for any function $f \in \mathcal{F}(\mathcal{H})$, so that

$$f(x) = \int_{\mathcal{H}} e^{\mathrm{i}(x,\alpha)} \mathrm{d}\mu(\alpha) \ , \tag{4.34}$$

the representation

$$f(x) = \int_{\mathcal{H}} e^{\mathrm{i}\langle x,\alpha\rangle} \mathrm{d}\nu(\alpha) \ , \tag{4.35}$$

if we take ν to be the measure defined by

$$\int_{\mathcal{H}} h(\alpha)\mathrm{d}\nu(\alpha) = \int_{\mathcal{H}} h(B^{-1}\alpha)\mathrm{d}\mu(\alpha) \ .$$

It follows now from (4.33) that

$$\widetilde{\int_{\mathcal{H}}} e^{\frac{i}{2}\langle x,x\rangle} f(x)\mathrm{d}x = \int_{\mathcal{H}} e^{-\frac{i}{2}\langle\alpha,\alpha\rangle} \mathrm{d}\nu(\alpha) \ . \tag{4.36}$$

From this we can see that it is natural to generalize the normalized integral on a Hilbert space treated in Chap. 2 to the following situation. Let E be a real separable Banach space on which we have a non degenerate bounded symmetric bilinear form $\langle x,y\rangle$, where non degenerate simply means that the continuous mapping from E into E^*, the dual of E, given by the form $\langle x,y\rangle$, is one to one. Let $\mathcal{F}(E,\langle\rangle)$ be the Banach space of continuous functions on E of the form

$$f(x) = \int_{E} e^{\mathrm{i}\langle x,\alpha\rangle} \mathrm{d}\mu(\alpha) \ , \tag{4.37}$$

where μ is a bounded complex measure on E with norm $||f||_0 = ||\mu||$ Since E is a separable metric group it easily follows as in section 2 that $\mathcal{F}(E, \langle \, \rangle)$ is a Banach algebra. We now define the normalized integral on E, equipped with the non degenerate form $\langle \, \rangle$, by

$$\overset{\sim}{\int_E} e^{\frac{i}{2}\langle x,x\rangle} f(x)\mathrm{d}x = \int_E e^{-\frac{i}{2}\langle \alpha,\alpha\rangle}\mathrm{d}\mu(\alpha) \,, \qquad (4.38)$$

and from (4.36) we have that in the case where E is a separable Hilbert space and $\langle x, y \rangle = (x, By)$, where B and B^{-1} are both bounded and everywhere defined, the normalized integral (4.38) is the same as the integral normalized with respect to the form $(x, B^{-1}x)$. It follows easily as in section 2 that (4.38) is a bounded continuous linear functional on $\mathcal{F}(E, \langle \, \rangle)$.

Proposition 4.3. *Let E_1 and E_2 be two real separable Banach spaces and let $\langle x_1, x_2 \rangle$ be a non degenerate bounded symmetric bilinear form on E_1. Let T be a bounded one-to-one mapping of E_2 into E_1 with a bounded inverse. Then $\langle Ty_1, Ty_2 \rangle$ is a non degenerate bounded symmetric bilinear form on E_2. Moreover if $f \in \mathcal{F}(E_1, \langle, \rangle)$ then $f(Ty)$ is in $\mathcal{F}(E_2, \langle T\cdot, T\cdot \rangle)$ and*

$$\overset{\sim}{\int_{E_1}} e^{\frac{i}{2}\langle x,x\rangle} f(x)\mathrm{d}x = \overset{\sim}{\int_{E_2}} e^{\frac{i}{2}\langle Ty,Ty\rangle} f(Ty)\mathrm{d}y \,.$$

Proof. The non degeneracy of $\langle Ty_1, Ty_2 \rangle$ follows from the fact that $\langle x_1, x_2 \rangle$ is non degenerate and that T is one-to-one continuous and with a range equal to E_1. Let now

$$f(x) = \int_{E_1} e^{i\langle x,\alpha\rangle}\mathrm{d}\mu(\alpha) \,,$$

then

$$f(Ty) = \int_{E_1} e^{i\langle Ty,\alpha\rangle}\mathrm{d}\mu(\alpha)$$

$$= \int_{E_2} e^{i\langle Ty,T\beta\rangle}\mathrm{d}\nu(\beta) \,,$$

where ν is the measure on E_2 induced by μ and the continuous transformation T^{-1} from E_1 to E_2. Hence $f(Ty)$ is in $\mathcal{F}(E_2, \langle T\cdot, T\cdot \rangle)$ and

$$\overset{\sim}{\int_{E_2}} e^{\frac{i}{2}\langle Ty,Ty\rangle} f(Ty)\mathrm{d}y = \int_{E_2} e^{-\frac{i}{2}\langle T\beta,T\beta\rangle}\mathrm{d}\nu(\beta)$$

$$= \int_{E_1} e^{-\frac{i}{2}\langle \alpha,\alpha\rangle}\mathrm{d}\mu(\alpha) \,.$$

This then proves the proposition. $\qquad\qquad\square$

Proposition 4.4. *Let E be a real separable Banach space with a bounded symmetric non-degenerate bilinear form $\langle x, y \rangle$. Let $E = E_1 \oplus E_2$ be a splitting of E into two closed subspaces E_1 and E_2 such that $\langle x_1, x_2 \rangle = 0$ for $x_1 \in E_1$ and $x_2 \in E_2$. Assume now that the restriction of the form $\langle x, y \rangle$ to $E_1 \times E_1$ is non degenerate, then the restriction to $E_2 \times E_2$ is also non degenerate and for any $f \in \mathcal{F}(E, \langle, \rangle)$ we have*

$$\int_E^{\sim} e^{\frac{i}{2}\langle x,x \rangle} f(x) dx = \int_{E_1}^{\sim} e^{\frac{i}{2}\langle x_1,x_1 \rangle} \left[\int_{E_2}^{\sim} e^{\frac{i}{2}\langle x_2,x_2 \rangle} f(x_1, x_2) dx_2 \right] dx_1 \;,$$

with $f(x_1, x_2) = f(x_1 \oplus x_2)$ for $x_1 \in E_1$ and $x_2 \in E_2$.

Proof. Let $x_2 \in E_2$ and assume that $\langle x_2, y_2 \rangle = 0$ for all $y_2 \in E_2$. Since $E = E_1 \times E_2$ we then have that $\langle x_2, y \rangle = 0$ for all $y \in E$, hence $x_2 = 0$, and the form restricted to $E_2 \times E_2$ is non degenerate. Since now $E = E_1 \oplus E_2$, E is equivalent as a metric group with $E_1 \times E_2$ and therefore

$$f(x) = \int_{E_1 \times E_2} e^{i\langle x, \alpha_1 + \alpha_2 \rangle} d\mu(\alpha_1, \alpha_2) \;,$$

hence

$$f(x_1, x_2) = \int e^{i\langle x_1, \alpha_1 \rangle} \cdot e^{i\langle x_2, \alpha_2 \rangle} d\mu(\alpha_1, \alpha_2) \;.$$

So that, for fixed x_1, $f(x_1, x_2) \in \mathcal{F}(E_2, \langle, \rangle)$ and

$$\int_{E_2}^{\sim} e^{\frac{i}{2}\langle x_2,x_2 \rangle} f(x_1, x_2) dx_2 = \int_{E_2 \times E_1} e^{-\frac{i}{2}\langle \alpha_2, \alpha_2 \rangle} e^{i\langle x_1, \alpha_1 \rangle} d\mu(\alpha_1, \alpha_2) \;. \quad (4.39)$$

We see that (4.39) is in $\mathcal{F}(E_1, \langle \rangle)$ and we compute

$$\int_{E_1}^{\sim} e^{\frac{i}{2}\langle x_1,x_1 \rangle} \left[\int_{E_2}^{\sim} e^{\frac{i}{2}\langle x_2,x_2 \rangle} f(x_1, x_2) dx_2 \right] dx_1$$

$$= \int_{E_1 \times E_2} e^{-\frac{i}{2}\langle \alpha_1, \alpha_1 \rangle} e^{-\frac{i}{2}\langle \alpha_2, \alpha_2 \rangle} d\mu(\alpha_1, \alpha_2)$$

$$= \int_E e^{-\frac{i}{2}\langle \alpha, \alpha \rangle} d\mu(\alpha) \;,$$

and this proves the proposition.   □

In the case where E is a Hilbert space and one has $\langle x, y \rangle = (x, By)$, with B and B^{-1} both bounded, a stronger version of Proposition 4.4 holds, we have namely:

Proposition 4.5. *Let $\mathcal{H}$ be a real separable Hilbert space and let $\langle x, y \rangle = (x, By)$, with B symmetric, such that $D(B) = R(B) = \mathcal{H}$,[2] with a bounded inverse B^{-1}, then $\mathcal{F}(\mathcal{H}, \langle \rangle) = \mathcal{F}(\mathcal{H})$. If $\mathcal{H}_1$ is a closed subspace of $\mathcal{H}$ such that the restriction of $\langle x, y \rangle$ to $\mathcal{H}_1 \times \mathcal{H}_1$ is non degenerate, and $\mathcal{H}_2 = B^{-1}\mathcal{H}_1^\perp$, where $\mathcal{H}_1^\perp$ is the orthogonal complement of $\mathcal{H}_1$ in $\mathcal{H}$, then $\mathcal{H} = \mathcal{H}_1 \oplus \mathcal{H}_2$ is a splitting of $\mathcal{H}$ in two closed subspaces such that the restriction of $\langle x, y \rangle$ to $\mathcal{H}_1 \times \mathcal{H}_2$ is identically zero. Hence by the previous proposition the restriction of $\langle x, y \rangle$ to $\mathcal{H}_2 \times \mathcal{H}_2$ is non degenerate and for any $f \in \mathcal{F}(\mathcal{H})$ we have*

$$\widetilde{\int_{\mathcal{H}}} e^{\frac{i}{2}\langle x,x \rangle} f(x)\mathrm{d}x = \widetilde{\int_{\mathcal{H}_1}} e^{\frac{i}{2}\langle x_1,x_1 \rangle} \left[\widetilde{\int_{\mathcal{H}_2}} e^{\frac{i}{2}\langle x_2,x_2 \rangle} f(x_1, x_2)\mathrm{d}x_2 \right] \mathrm{d}x_1 ,$$

with $f(x_1, x_2) = f(x_1 \oplus x_2)$, where $x = x_1 \oplus x_2$ is the splitting of $\mathcal{H}$ into $\mathcal{H}_1 \oplus \mathcal{H}_2$ and the sum is orthogonal with respect to $\langle x, y \rangle = (x, By)$.

Proof. That $\mathcal{F}(\mathcal{H}, \langle \rangle) = \mathcal{F}(\mathcal{H})$ was already proved by (4.34) and (4.35). So let now $\mathcal{H}_1$ and $\mathcal{H}_2$ be as in the proposition. Since B^{-1} has a bounded inverse B, $\mathcal{H}_2 = B^{-1}\mathcal{H}_1^\perp$ is obviously a closed subspace and $\langle x, y \rangle = 0$ for $x \in \mathcal{H}_1$ and $y \in \mathcal{H}_2$. Let $x \in \mathcal{H}$ be such that $\langle x, y \rangle = 0$ for all $y \in \mathcal{H}_1 + \mathcal{H}_2$. Then $\langle x, y \rangle = 0$ for all $y \in \mathcal{H}_1$, hence $x \in \mathcal{H}_2$. Now let $y \in \mathcal{H}_2$, then $y = B^{-1}z$ with $z \in \mathcal{H}_1^\perp$ and $0 = \langle x, y \rangle = (x, z)$ for all $z \in \mathcal{H}_1^\perp$, so that $x \in \mathcal{H}_1$, but since $x \in \mathcal{H}_2$ we have also $\langle x, y \rangle = 0$ for all y in $\mathcal{H}_1$. Therefore, by the non degeneracy of $\langle x, y \rangle$ in $\mathcal{H}_1 \times \mathcal{H}_1$, we have that $x = 0$. This shows that $\mathcal{H}_1 \cap \mathcal{H}_2 = \{0\}$ and that the orthogonal complement of $B(\mathcal{H}_1 + \mathcal{H}_2)$ is zero. Hence $B(\mathcal{H}_1 + \mathcal{H}_2)$ is dense and since B and B^{-1} both are bounded and therefore preserve the topology, we have that $\mathcal{H}_1 + \mathcal{H}_2$ is dense in $\mathcal{H}$. But since now both $\mathcal{H}_1$ and $\mathcal{H}_2$ are closed and $\mathcal{H}_1 \cap \mathcal{H}_2 = \{0\}$ we get that $\mathcal{H}_1 + \mathcal{H}_2 = \mathcal{H}$. This proves the proposition. $\square$

Notes

This general approach was first presented in the first edition of this book. Special cases have been presented later on in several contexts, under more restrictive assumptions on the phase function, but going deeper in the analysis and leading to applications like the treatment of magnetic fields [70] (see Sect. 10.5.1).

---

[2] From Hellinger–Toeplitz theorem (see e.g. [56]) it follows then that B is necessarily bounded

5

Feynman Path Integrals
for the Anharmonic Oscillator

By the anharmonic oscillator with n degrees of freedom we shall understand the mechanical system in $\mathbb{R}^n$ with classical action integral of the form:

$$S_t = \frac{m}{2} \int_0^t \dot{\gamma}(\tau)^2 d\tau - \frac{1}{2} \int_0^t \gamma A^2 \gamma d\tau - \int_0^t V(\gamma(\tau)) d\tau , \qquad (5.1)$$

where A^2 is a strictly positive definite matrix in $\mathbb{R}^n$ and $\dot{\gamma}(\tau) = \frac{d\gamma}{d\tau}$ and $V(x)$ is a nice function which in the following shall be taken to be in the space $\mathcal{F}(\mathbb{R}^n)$ i.e.

$$V(x) = \int_{\mathbb{R}^n} e^{i\alpha x} d\mu(\alpha) , \qquad (5.2)$$

where μ is a bounded complex measure. We shall of course also assume, for physical reasons, that V is real[1]. Let $\varphi(x) \in \mathcal{F}(\mathbb{R}^n)$ with

$$\varphi(x) = \int_{\mathbb{R}^n} e^{i\alpha x} d\nu(\alpha) , \qquad (5.3)$$

then we shall give a meaning to the Feynman path integral

$$\widetilde{\int_{\gamma(t)=x}} e^{\frac{1}{2}\left(\int_0^t \dot{\gamma}^2 d\tau - \int_0^t \gamma A^2 \gamma d\tau\right)} \cdot e^{-i \int_0^t V(\gamma(\tau)) d\tau} \varphi(\gamma(0)) d\gamma , \qquad (5.4)$$

by using the integral defined in the previous section. For simplicity we shall assume in what follows that $m = \hbar = 1$. In the previous section we only defined integrals over linear spaces, so we shall first have to transform the

---

[1] This condition is however not necessary for the mathematics involved, so that all results actually hold also for complex V.

non homogeneous boundary condition $\gamma(t) = x$ into a homogeneous one. This is easily done if there exists a solution $\beta(\tau)$ to the following boundary value problem on the interval $[0, t]$:

$$\ddot{\beta} + A^2\beta = 0 , \quad \beta(t) = x , \quad \dot{\beta}(0) = 0 . \tag{5.5}$$

Let $\lambda_1, \ldots, \lambda_n$ be the eigenvalues of A. If we now assume that

$$t \neq \left(k + \frac{1}{2}\right)\frac{\pi}{\lambda_i} \tag{5.6}$$

for all $k = 0, 1, \ldots$, and $i = 1, \ldots, n$, then (5.5) has a unique solution given by

$$\beta(\tau) = \frac{\cos A\tau}{\cos At}x . \tag{5.7}$$

We then make formally the substitution $\gamma \to \gamma + \beta$ in (5.4) and get

$$\int\limits_{\gamma(t)=0} e^{\frac{i}{2}\int_0^t (\dot{\gamma}+\dot{\beta})^2 d\tau - \frac{1}{2}\int_0^t (\gamma+\beta)A^2(\gamma+\beta)d\tau} e^{-i\int_0^t V(\gamma(\tau)+\beta(\tau))d\tau} \varphi(\gamma(0) + \beta(0))d\gamma .$$

$$\tag{5.8}$$

Now, due to (5.5), we have that, if $\gamma(t) = 0$,

$$\int\limits_0^t (\dot{\gamma} + \dot{\beta})^2 d\tau - \int\limits_0^t (\gamma + \beta)A^2(\gamma + \beta)d\tau$$

$$= \int\limits_0^t \dot{\gamma}^2 d\tau - \int\limits_0^t \gamma A^2\gamma d\tau + \beta(t)\dot{\beta}(t) . \tag{5.9}$$

Since $\beta(t) = x$ and $\dot{\beta}(t) = -A\, tg\, At\, x$, we may write (5.8) as

$$e^{-\frac{i}{2}xA\, tg\, A\, x} \int\limits_{\gamma(t)=0} e^{\frac{i}{2}\int_0^t (\dot{\gamma}^2 - \gamma A^2\gamma)d\tau} e^{-i\int_0^t V(\gamma(\tau)+\beta(\tau))d\tau} \varphi(\gamma(0) + \beta(0))d\gamma .$$

$$\tag{5.10}$$

Hence we have transformed the boundary condition to a homogeneous one. Let now $\mathcal{H}_0$ be the real separable Hilbert space of continuous functions γ from $[0, t]$ to $\mathbb{R}^n$ such that $\gamma(t) = 0$ and $|\gamma|^2 = \int_0^t \dot{\gamma}^2 d\tau$ is finite. In $\mathcal{H}_0$ the quadratic form $\int_0^t (\dot{\gamma}^2 - \gamma A^2\gamma)d\tau$ is obviously bounded and therefore given by a bounded symmetric operator B in $\mathcal{H}_0$, so that, with $(\gamma, \gamma) = |\gamma|^2$, we have

$$(\gamma, B\gamma) = \int\limits_0^t (\dot{\gamma}^2 - \gamma A^2\gamma)d\tau . \tag{5.11}$$

From the Sturm–Liouville theory we also have that B is onto with a bounded inverse B^{-1}, if $\lambda = 0$ is not an eigenvalue of, the following eigenvalue problem on $[0, t]$:

$$\ddot{u} + A^2 u = \lambda \ddot{u}, \quad u(t) = 0, \quad \dot{u}(0) = 0. \tag{5.12}$$

One easily verifies that if t satisfies (5.6), then zero is not an eigenvalue for (5.12) and the Green's function for the eigenvalue problem (5.12) is given by

$$g_0(\sigma, \tau) = \frac{\cos A\sigma \sin A(t - \tau)}{A \cos At} \quad \text{for} \quad \sigma \leq \tau. \tag{5.13}$$

We shall now assume that t satisfies (5.6) i.e. that $\cos At$ is non degenerate. Then since the range of B is equal to all $\mathcal{H}_0$ and B^{-1} is bounded we are in the case where the space D is equal to $\mathcal{H}_0$ and $\triangle$ is uniquely given by $(\gamma, B^{-1}\gamma)$. Using then the notation introduced in (4.33) with $\langle \gamma_1, \gamma_2 \rangle = (\gamma_1, B\gamma_2)$ i.e.

$$\langle \gamma_1, \gamma_2 \rangle = \int_0^t (\dot{\gamma}_1 \dot{\gamma}_2 - \gamma_1 A^2 \gamma_2) d\tau, \tag{5.14}$$

we verify exactly as in Chap. 3 that

$$f(\gamma) = e^{-i \int_0^t V(\gamma(\tau) + \beta(\tau)) d\tau} \varphi(\gamma(0) + \beta(0))$$

is in $\mathcal{F}(\mathcal{H}_0)$. Hence

$$e^{-\frac{i}{2}xA \, tg \, tAx} \int_{\mathcal{H}_0} e^{\frac{i}{2}\langle \gamma, \gamma \rangle} f(\gamma) d\gamma \tag{5.15}$$

is well defined and we take (5.15) to be the definition of the Feynman path integral (5.4), which we shall now compute. As in Chap. 3 we have that

$$f(\gamma) = \sum_{n=0}^{\infty} (-i)^n \int \cdots \int_{0 \leq t_1 \leq \ldots \leq t_n \leq t} \int \cdots \int$$

$$e^{i \sum_{j=0}^n \alpha_j \beta(t_j)} e^{i \sum_{j=0}^n \alpha_j \gamma(t_j)} d\nu(\alpha_0) \prod_{j=1}^n d\mu(\alpha_j) dt_j \tag{5.16}$$

where $t_0 = 0$, the sum converges strongly in $\mathcal{F}(\mathcal{H}_0)$ and the integrands are continuous in α_j and t_j in the weak $\mathcal{F}(\mathcal{H}_0)$ topology, hence weakly and therefore strongly integrable in $\mathcal{F}(\mathcal{H}_0)$, by the theorem of Pettis ([52], p. 131).

Since $\mathcal{F}_\triangle$ is continuous in $\mathcal{F}(D^*) = \mathcal{F}(\mathcal{H}_0)$ we may therefore commute the sum and the integrals in (5.16) with the integral in (5.15). Hence it suffices to compute

$$\int_{\mathcal{H}_0} e^{\frac{i}{2}\langle \gamma, \gamma \rangle} e^{i \sum \alpha_j \gamma(t_j)} d\gamma. \tag{5.17}$$

Now $\gamma_i(\sigma) \in \mathcal{H}_0^* = \mathcal{H}_0$, hence it is of the form $\gamma_i(\sigma) = \langle \gamma, \gamma_\sigma^i \rangle$ and therefore, by the Definition 5.17 is equal to

$$e^{-\frac{1}{2} \sum_{ij} \alpha_i \langle \gamma_{t_i}, \gamma_{t_j} \rangle \alpha_j} . \tag{5.18}$$

Using now that (5.13) is the Green's function for (5.12) one verifies by standard computations that

$$\langle \gamma_\sigma, \gamma_\tau \rangle = \frac{\sin A(t - \sigma \vee \tau) \cos A(\sigma \wedge \tau)}{A \cos At} . \tag{5.19}$$

So we have computed (5.15) and shown that it is equal to

$$e^{-\frac{1}{2} xA \, tg \, tA \, x} \sum_{n=0}^\infty (-i)^n \int_{0 \le t_1 \le \ldots \le t_n \le t} \cdots \int \int \cdots \int$$

$$e^{i \sum_{j=0}^n \alpha_j \beta(t_j)} \; e^{-\frac{i}{2} \sum_{jk=0}^n \alpha_j \langle \gamma_{t_j}, \gamma_{t_k} \rangle \alpha_k} \, d\nu(\alpha_0) \prod_{j=1}^n d\mu(\alpha_j) dt_j$$

$$= e^{-\frac{1}{2} xA \, tg \, tA \, x} \sum_{n=0}^\infty (-i)^n \int_{0 \le t_1 \ldots \le t} \cdots \int \int \int \cdots \int$$

$$e^{i \sum_{j=0}^n \alpha_j \frac{\cos At_j}{\cos At} x} \; e^{-\frac{i}{2} \sum_{jk=0}^n \alpha_j \langle \gamma_{t_j}, \gamma_{t_k} \rangle \alpha_k} \, d\nu(\alpha_0) \prod_{j=1}^n d\mu(\alpha_j) dt_j . \tag{5.20}$$

Let us now define $\Omega_0(x)$ by

$$\Omega_0(x) = \left| \frac{1}{\pi} A \right|^{1/4} e^{-\frac{1}{2} xAx} , \tag{5.21}$$

where $\left| \frac{1}{\pi} A \right|$ is the determinant of the transformation $\frac{1}{\pi} A$. It is well known Ω_0 is the normalized eigenfunction for the lowest eigenvalue of the self adjoint operator

$$H_0 = -\frac{1}{2} \Delta + \frac{1}{2} xA^2 x \tag{5.22}$$

in $L_2(\mathbb{R}^n)$. We have in fact

$$H_0 \Omega_0 = \frac{1}{2} \mathrm{tr} \, A \Omega_0 , \tag{5.23}$$

where $\mathrm{tr} \, A$ is the trace of A.

For $0 \le t_1 \le \ldots \le t_n \le t$ we shall now compute the Feynman path integral

$$I(x) = \int\limits_{\gamma(t)=x} e^{\frac{i}{2}\left(\int_0^t \dot{\gamma}^2(\tau)d\tau - \int_0^t \gamma A^2 \gamma d\tau\right)} e^{i \sum\limits_{j=1}^n \alpha_j \gamma(t_j)} \Omega_0(\gamma(0))d\gamma$$

$$\stackrel{\text{Def}}{=} \int\limits_{\mathcal{H}_0} e^{\frac{i}{2}\int_0^t (\dot{\gamma}+\dot{\beta})^2 d\tau - \frac{i}{2}\int_0^t (\gamma+\beta)A^2(\gamma+\beta)d\tau} e^{i \sum\limits_{j=1}^n \alpha_j(\gamma(t_j)+\beta(t_j))}$$

$$\Omega_0\left(\gamma(0) + \beta(0)\right) d\gamma . \tag{5.24}$$

By the previous calculation we have

$$I(x) = e^{-\frac{i}{2}xA \, \text{tg} \, t \, Ax} \int\limits_{\mathbb{R}^n} e^{i \sum\limits_{j=0}^n \alpha_j \frac{\cos A \, t_j}{\cos A \, t}x} e^{-\frac{i}{2} \sum\limits_{jk=0}^n \alpha_j \langle \gamma_{t_j}, \gamma_{t_k} \rangle \alpha_k} d\mu_0(\alpha_0) , \tag{5.25}$$

where $t_0 = 0$ and $d\nu_0(\alpha_0)$ is given by

$$\Omega_0(x) = \int\limits_{\mathbb{R}^n} e^{ix\alpha_0} d\nu_0(\alpha_0) , \tag{5.26}$$

from which we get

$$d\nu_0(\alpha_0) = |4\pi^3 A|^{-1/4} e^{-\frac{1}{2}\alpha_0 A^{-1}\alpha_0} d\alpha_0 . \tag{5.27}$$

Substituting (5.27) in (5.25) we obtain

$$I(x) = e^{-\frac{i}{2}xA \, \text{tg} \, tAx} e^{i \sum\limits_{j=1}^n \alpha_j \frac{\cos A \, t_j}{\cos A \, t}x} e^{-\frac{i}{2}\sum\limits_{jk=1}^n \alpha_j \langle \gamma_{t_j},\gamma_{t_k}\rangle\alpha_k}$$

$$|4\pi^3 A|^{-1/4} \int\limits_{\mathbb{R}^n} e^{i\alpha_0 \frac{1}{\cos A \, t}x} e^{-i\alpha_0 \sum\limits_{j=1}^n g_0(0,t_j)\alpha_j} e^{-\frac{1}{2}\alpha_0 g_0(0,0)\alpha_0}$$

$$\cdot \, e^{-\frac{1}{2}\alpha_0 A^{-1}\alpha_0} d\alpha_0 . \tag{5.28}$$

Now the integral above is equal to

$$\left|2\pi A \cos tA \, e^{-itA}\right|^{\frac{1}{2}} e^{-\frac{1}{2}xA(1-i \tan tA)x} e^{x \sum\limits_{j=1}^n \frac{\sin A(t-t_j)e^{-itA}}{\cos At}\alpha_j}$$

$$e^{-\frac{1}{2}\sum\limits_{jk=1}^n \alpha_j \frac{\sin A(t-t_j)\sin A(t-t_k)}{A \cos At \, e^{itA}}\alpha_k} . \tag{5.29}$$

Hence we have finally that[2]

$$I(x) = \left|\frac{1}{\sqrt{\pi}}A^{\frac{1}{2}}\cos tA\right|^{\frac{1}{2}} e^{-\frac{it}{2}tr A} e^{-\frac{i}{2}xAx} e^{ix \sum\limits_{j=1}^n e^{-i(t-t_j)A}\alpha_j}$$

$$\cdot \, e^{-\frac{i}{2}\sum\limits_{jk=1}^n \alpha_j A^{-1}e^{-i(t-t_j \wedge t_k)A}\sin A(t-t_j \vee t_k)\alpha_k} . \tag{5.30}$$

---

[2] In the first edition of this book (5.30) contained a misprint. We present here the exact formula, taking into account a suggestion by P.G. Hjiorth, to whom we are very grateful.

Now by an explicit calculation we obtain from (5.30) that

$$\int I(x)\Omega_0(x)\mathrm{d}x = |\cos tA|^{\frac{1}{2}}\mathrm{e}^{-\frac{\mathrm{i}t}{2}\mathrm{tr}\,A}\mathrm{e}^{-\frac{1}{2}\sum\limits_{j k=1}^{n}\alpha_j(2A)^{-1}\mathrm{e}^{-\mathrm{i}|t_k-t_j|A}\alpha_k}. \tag{5.31}$$

On the other hand it is well known from the standard theory of the quantum mechanics for the harmonic oscillator that, for $0 \le t_1 \le \ldots \le t_n \le t$,

$$\left(\Omega_0, \mathrm{e}^{\mathrm{i}\alpha_1 x(t_1)}\ldots\mathrm{e}^{\mathrm{i}\alpha_n x(t_n)}\mathrm{e}^{-\mathrm{i}tH_0}\Omega_0\right)$$

$$= \mathrm{e}^{-\frac{\mathrm{i}t}{2}\mathrm{tr}\,A}\mathrm{e}^{-\frac{1}{2}\sum\limits_{j,k=1}^{n}\alpha_j(2A)^{-1}\mathrm{e}^{-\mathrm{i}|t_j-t_k|A}\alpha_k}, \tag{5.32}$$

where $\mathrm{e}^{\mathrm{i}\alpha x(\tau)} = \mathrm{e}^{-\mathrm{i}\tau H_0}\mathrm{e}^{\mathrm{i}\alpha x}\mathrm{e}^{\mathrm{i}\tau H_0}$. Hence we have proved the formula

$$|\cos At|^{-\frac{1}{2}}\int_{\mathbb{R}^n}\Omega_0(x)\left[\int_{\gamma(t)=x}^{\sim}\mathrm{e}^{\frac{1}{2}\int_0^t(\dot\gamma^2-\gamma A^2\gamma)\mathrm{d}\tau}\mathrm{e}^{\mathrm{i}\sum\limits_{j=1}^{n}\alpha_j\gamma(t_j)}\Omega_0(\gamma(0))\mathrm{d}\gamma\right]\mathrm{d}x$$

$$= \left(\Omega_0, \mathrm{e}^{\mathrm{i}\alpha_1 x(t_1)}\ldots\mathrm{e}^{\mathrm{i}\alpha_n x(t_n)}\mathrm{e}^{-\mathrm{i}tH_0}\Omega_0\right) \tag{5.33}$$

for $0 \le t_1 \le \ldots \le t_n \le t$.[3]

Let now

$$H = H_0 + V(x). \tag{5.34}$$

We have the norm convergent expansion

$$\mathrm{e}^{-\mathrm{i}tH} = \sum_{n=0}^{\infty}(-\mathrm{i})^n\int\cdots\int_{0\le t_1\le\ldots\le t_n\le t}V(t_1)\ldots V(t_n)\mathrm{e}^{-\mathrm{i}tH_0}\mathrm{d}t_1\ldots\mathrm{d}t_n, \tag{5.35}$$

where $V(\tau) = \mathrm{e}^{-\mathrm{i}\tau H_0}V\mathrm{e}^{\mathrm{i}\tau H_0}$.

If now f, g and V are taken to be in $\mathcal{F}(\mathbb{R}^n)$, then we get from (5.35), (5.34) and the fact that the sum and the integrals in (5.16) can be taken in the strong sense in $\mathcal{F}(\mathcal{H}_0)$, that

$$|\cos At|^{\frac{1}{2}}\int_{\mathbb{R}^n}f(x)\Omega_0(x)\left[\int_{\gamma(t)=x}^{\sim}\mathrm{e}^{\frac{1}{2}\int_0^t(\dot\gamma^2-\gamma A^2\gamma)\mathrm{d}\tau}\mathrm{e}^{-\mathrm{i}\int_0^t V(\gamma(\tau))\mathrm{d}\tau}\right.$$

$$\left. g(\gamma(0))\Omega_0(\gamma(0))\mathrm{d}\gamma\right]\mathrm{d}x$$

$$= (\Omega_0, f\mathrm{e}^{-\mathrm{i}tH}g\Omega_0). \tag{5.36}$$

----

[3] This formula will be rewritten in another form in (6.21) below.

This then, by the density of $\mathcal{F}(\mathbb{R}^n)\Omega_0$ in $L_2(\mathbb{R}^n)$, proves the formula

$$\psi(x,t) = |\cos At|^{-\frac{1}{2}} \int\limits_{\gamma(t)=x} e^{\frac{i}{2}\int_0^t (\dot{\gamma}^2 - \gamma A^2 \gamma)d\tau} \, e^{-i\int_0^t V(\gamma(\tau))d\tau} \, \varphi(\gamma(0))d\gamma \quad (5.37)$$

for the solution of the Schrödinger equation for the anharmonic oscillator

$$i\frac{\partial}{\partial t}\psi = -\frac{1}{2}\triangle\psi + \frac{1}{2}xA^2x\psi + V(x)\psi \quad (5.38)$$

for all t such that $\cos At$ is non singular and $V \in \mathcal{F}(\mathbb{R}^n)$ and the initial condition $\psi(x,0) = \varphi(x)$ is in $\mathcal{F}(\mathbb{R}^n) \cap L_2(\mathbb{R}^n)$. From (5.36) we only get (5.37) for $\varphi \in \mathcal{F}(\mathbb{R}^n) \cdot \Omega_0$, but since the left hand side of (5.37) is continuous in L_2 as a function of φ and the right hand side for fixed x is continuous as a function of φ in $\mathcal{F}(\mathbb{R}^n)$, by the fact that the sum and the integrals in (5.16) can be taken in the strong $\mathcal{F}(\mathcal{H}_0)$ sense, we get (5.37) for all $\varphi \in \mathcal{F}(\mathbb{R}^n) \cap L_2(\mathbb{R}^n)$. Although the integral over γ in (5.37) was defined by (5.15) using the translation by β, we have, since $D(B) = \mathcal{H}_0$, by Proposition 4.2, that the integral in (5.15) is invariant under translations by any $\gamma_0 \in \mathcal{H}_0$ i.e. by any path $\gamma_0(t)$ for which $\gamma_0(t) = 0$ and the kinetic energy $\frac{1}{2}\int_0^t \dot{\gamma}_0(\tau)^2 d\tau$ is finite. Hence as a matter of fact the definition of the integral over γ in (5.37) does not depend on the specific choice of β. We state these results, for the case when m and $\hbar$ are not necessarily both equal to one, in the following theorem.

Theorem 5.1. *Let $\mathcal{H}_0$ be the real separable Hilbert space of continuous functions γ from $[0,t]$ into $\mathbb{R}^n$ such that $\gamma(t) = 0$ and with finite kinetic energy $\frac{m}{2}\int_0^t \dot{\gamma}(\tau)^2 d\tau$, and norm given by $|\gamma|^2 = \int_0^t \dot{\gamma}(\tau)^2 d\tau$. Let B be the bounded symmetric operator on $\mathcal{H}_0$, with $D(B) = \mathcal{H}_0$, given by $(\gamma_1, B\gamma_2) = \langle \gamma_1, \gamma_2 \rangle$ with*

$$\langle \gamma, \gamma \rangle = \frac{1}{\hbar}\int\limits_0^t (m\dot{\gamma}(\tau)^2 - \gamma A^2 \gamma)d\gamma ,$$

where A^2 is a strictly positive definite matrix in $\mathbb{R}^n$. Then for all values of t such that $\cos At$ is a non singular transformation in $\mathbb{R}^n$ we have that the range of B is $\mathcal{H}_0$ and B^{-1} is a bounded symmetric operator on $\mathcal{H}_0$. Hence, by Proposition 4.2, $D = \mathcal{H}_0$ and $\triangle(x,y)$ is uniquely given by B. Let $\beta(\tau)$ be any continuous path with $\beta(t) = x$ and finite kinetic energy, and let V and φ be in $\mathcal{F}(\mathbb{R}^n)$, then

$$f(\gamma) = e^{\frac{i}{2}\langle \beta, \beta \rangle} e^{i\langle \gamma, \beta \rangle} \cdot e^{-i\int_0^t V(\gamma(\tau) + \beta(\tau))d\tau} \varphi(\gamma(0) + \beta(0))$$

is in $\mathcal{F}(\mathcal{H}_0)$ and

$$\int\limits_{\mathcal{H}_0} e^{\frac{i}{2}\langle \gamma, \gamma \rangle} f(\gamma)d\gamma \quad (5.39)$$

does not depend on β. Moreover if φ is also an $L_2(\mathbb{R}^n)$ then

$$\psi(x,t) = |\cos At|^{-\frac{1}{2}} \int\limits_{\gamma(t)=x} e^{\frac{1}{2\hbar}\int\limits_0^t (m\dot{\gamma}^2 - \gamma A^2 \gamma)d\tau} \, e^{-\frac{i}{\hbar}\int\limits_0^t V(\gamma(\tau))d\tau} \, \varphi(\gamma(0))d\gamma \,,$$

where the integral over γ is defined by (5.39), and $\psi(x,t)$ is the solution of the Schrödinger equation for the anharmonic oscillator

$$i\hbar\frac{\partial}{\partial t}\psi(x,t) = -\frac{\hbar^2}{2m}\triangle\psi + \frac{1}{2}xA^2x\psi + V(x)\psi$$

with initial values $\psi(x,0) = \varphi(x)$.[4]

Let us now set, for $0 \le t_1 \le \ldots \le t_n \le t$, $\varphi(x) \in \mathcal{S}(\mathbb{R}^n)$ and t such that $\cos At$ is non singular:

$$I(x) = \int\limits_{\gamma(t)=x} e^{\frac{1}{2}\int\limits_0^t (\dot{\gamma}(\tau)^2 - \gamma A^2 \gamma)d\tau} \, e^{i\sum\limits_{j=1}^n \alpha_j\gamma(t_j)} \, \varphi(\gamma(0))d\gamma \qquad (5.40)$$

$$\stackrel{\text{Def}}{=} \int\limits_{\mathcal{H}_0}^{\triangle} e^{\frac{1}{2}\int\limits_{-t}^0 ((\dot{\gamma}+\dot{\beta})^2 - (\gamma+\beta)A^2(\gamma+\beta))d\tau} \, e^{i\sum\limits_{j=1}^n \alpha_j(\gamma(t_j)+\beta(t_j))} \, \varphi(\gamma(0)+\beta(0))d\gamma \,.$$

with $\triangle$ and $\mathcal{H}_0$ as given in the previous theorem, and $\beta(\tau)$ same path with finite kinetic energy such that $\beta(t) = x$. We then get in the same way as for (5.24) that

$$I(x) = e^{-\frac{1}{2}xA\,\mathrm{tg}\,tAx} \int\limits_{\mathbb{R}^n} e^{i\sum\limits_{j=0}^n \alpha_j \frac{\cos At_j}{\cos At}x} \, e^{-\frac{i}{2}\sum\limits_{jk=0}^n \alpha_j g_0(t_j,t_k)\alpha_k} \, \hat{\varphi}(\alpha_0)d\alpha_0 \qquad (5.41)$$

where $t_0 = 0$, $g_0(\sigma,\tau)$ is given by (5.13) and

$$\varphi(x) = \int\limits_{\mathbb{R}^n} e^{ix\alpha_0}\hat{\varphi}(\alpha_0)d\alpha_0 \,.$$

Since $\hat{\varphi} \in \mathcal{S}(\mathbb{R}^n)$ we get from (5.41) that $I(x) \in \mathcal{S}(\mathbb{R}^n)$, hence $I(x)$ is integrable and we have by direct computation, if $\sin At$ is non singular,

$$\int I(x)dx = \left|\frac{i}{2\pi}A\,\mathrm{tg}\,tA\right|^{-\frac{1}{2}} \int\limits_{\mathbb{R}^n} e^{\frac{i}{2}\sum\limits_{jk=0}^n \alpha_j \frac{\cos At_j \cos At_k}{A\sin At \cos At}\alpha_k}$$

$$e^{-\frac{i}{2}\sum\limits_{jk=0}^n \alpha_j g_0(t_j,t_k)\alpha_k} \, \hat{\varphi}(\alpha_0)d\alpha_0 \,.$$

---

[4] In Chaps. 5–7 all results are stated for the case of an anharmonic oscillator. From the proofs it is evident that they hold also for N anharmonic oscillators, with anharmonicity V which is a superposition of ν-body potentials ($\nu = 1, 2, \ldots$), which can also be translation invariant.

Hence we have

$$\int I(x)\mathrm{d}x = \left| \frac{\mathrm{i}}{2\pi} A \, tg \, tA \right|^{-\frac{1}{2}} \int_{\mathbb{R}^n} \mathrm{e}^{-\frac{\mathrm{i}}{2} \sum\limits_{jk=0}^{n} \alpha_j g(t_j,t_k)\alpha_k} \hat{\varphi}(\alpha_0)\mathrm{d}\alpha_0 \, , \qquad (5.42)$$

where $g(\sigma,\tau) = g(\tau,\sigma)$ and

$$g(\sigma,\tau) = -\frac{\cos A\sigma \cos A(t-\tau)}{A \sin At} \, . \qquad (5.43)$$

We now observe that $g(\sigma,\tau)$ is the Green's function for the self adjoint operator $-\frac{\mathrm{d}^2}{\mathrm{d}\tau^2} - A^2$ on $L_2([0,t];\mathbb{R}^n)$ with Neumann boundary conditions $\dot{\gamma}(0) = \dot{\gamma}(t) = 0$.

Let now $\mathcal{H}$ be the real separable Hilbert space of continuous functions γ from $[0,t]$ to $\mathbb{R}^n$ with finite kinetic energy without any conditions on the boundary and with norm given by

$$|\gamma|^2 = \int_0^t (\dot{\gamma}^2 + \gamma^2)\mathrm{d}\tau \, . \qquad (5.44)$$

Let B_N be the bounded symmetric operator with $D(B_N) = \mathcal{H}$, given by $(\gamma_1, B_N\gamma_2) = \langle\gamma_1,\gamma_2\rangle$ with

$$\langle\gamma,\gamma\rangle = \int_0^t (\dot{\gamma}^2 - \gamma A^2\gamma)\mathrm{d}\tau \, . \qquad (5.45)$$

Then if $\sin At$ is non singular we have from (5.43) that $R(B_N) = \mathcal{H}$ and B_N^{-1} is bounded. Hence with $D = B_N$ the form $\triangle_N$ of the previous section is uniquely given by B_N and $\triangle_N(\gamma_1,\gamma_2) = (\gamma_1, B_N^{-1}\gamma_2)$. Define now $\gamma_\sigma \in \mathcal{H} \times \mathbb{R}^n$ by

$$\langle\gamma,\gamma_\sigma\rangle = \gamma(\sigma) \, , \qquad (5.46)$$

then we get, by using the fact that $g(\sigma,\tau)$ is the Green's function for the Neumann boundary conditions, that

$$\langle\gamma_\sigma,\gamma_\tau\rangle = g(\sigma,\tau) \, . \qquad (5.47)$$

From this we obtain, for $0 \le t_1 \le \ldots \le t_n \le t$

$$\int_{\mathcal{H}} \mathrm{e}^{\frac{\mathrm{i}}{2}\langle\gamma,\gamma\rangle} \mathrm{e}^{\mathrm{i}\sum \alpha_j\gamma(t_j)}\mathrm{d}\gamma = \mathrm{e}^{-\frac{\mathrm{i}}{2} \sum\limits_{j,k=1}^{n} \alpha_j g(t_j,t_k)\alpha_k} \, . \qquad (5.48)$$

Let us now define for any $f(\gamma) \in \mathcal{F}(\mathcal{H})$ the path integral

$$\int_{\mathcal{H}}^{\sim} e^{\frac{i}{2}\int_0^t (\dot{\gamma}^2 - \gamma A^2 \gamma) d\tau} f(\gamma) d\gamma \overset{\text{Def}}{=} \int_{\mathcal{H}}^{\sim} e^{\frac{i}{2}\langle \gamma, \gamma \rangle} f(\gamma) d\gamma . \tag{5.49}$$

We then have that

$$\int_{\mathcal{H}}^{\sim} e^{\frac{i}{2}\int_0^t (\dot{\gamma}^2 - \gamma A^2 \gamma) d\tau} e^{i \sum \alpha_j \gamma(t_j)} d\gamma = e^{-\frac{i}{2} \sum\limits_{j,k=1}^{n} \alpha_j g(t_j, t_k) \alpha_k} . \tag{5.50}$$

Let now t be such that $\sin At$ and $\cos At$ both are non singular, and let us assume that $f(\gamma)$ is in $\mathcal{F}(\mathcal{H})$. Then one easily verifies that

$$e^{i\int_0^t (\dot{\gamma}\dot{\beta} - \gamma A^2 \beta) d\tau} f(\gamma + \beta) \tag{5.51}$$

is in $\mathcal{F}(\mathcal{H}_0)$ for any $\beta \in \mathcal{H}$, where $\mathcal{H}_0$ was defined as the space of paths γ with finite kinetic energy and such that $\gamma(t) = 0$. From (5.50) and (5.42) we now easily get the following formula, if we use the fact that the linear functionals $\gamma \to \gamma(\sigma)$ span a dense subset of $\mathcal{H}_0$ as well as of $\mathcal{H}$:

$$\left| \frac{i}{2\pi} A \, tg \, tA \right|^{\frac{1}{2}} \int_{\mathbb{R}^n} dx \int_{\gamma(t)=x}^{\sim} e^{\frac{i}{2}\int_0^t (\dot{\gamma}^2 - \gamma A^2 \gamma) d\tau} f(\gamma) d\gamma$$

$$= \int_{\mathcal{H}}^{\sim} e^{\frac{i}{2}\int_0^t (\dot{\gamma}^2 - \gamma A^2 \gamma) d\tau} f(\gamma) d\gamma . \tag{5.52}$$

By the same method as used in the proof of Theorem 5.1 we now have that, if φ_1, φ_2 and V are in $\mathcal{F}(\mathbb{R}^n)$, then

$$f(\gamma) = \overline{\varphi_1}(\gamma(t)) e^{-i\int_0^t V(\gamma(\tau)) d\tau} \varphi_2(\gamma(0)) \tag{5.53}$$

is in $\mathcal{F}(\mathcal{H})$, hence by (5.52) and Theorem 5.1 we get

$$(\varphi_1, e^{-itH}\varphi_2) = \left| \frac{i}{2\pi} A \sin At \right|^{-\frac{1}{2}} \int_{\mathcal{H}}^{\sim} e^{\frac{i}{2}\int_0^t (\dot{\gamma}^2 - \gamma A^2 \gamma) d\tau} e^{-i\int_0^t V(\gamma(\tau)) d\tau}$$

$$\overline{\varphi_1}(\gamma(t)) \varphi_2(\gamma(0)) d\gamma . \tag{5.54}$$

We have proved this formula only for values of t for which both $\cos At$ and $\sin At$ are non singular. However from (5.43) we see that, for fixed σ and τ, $g(\varphi, \tau)$ is a continuous function of t for values of t for which $\sin At$ is non singular. From this it easily follows that the right hand side of (5.54) is a continuous function of t for values of t for which $\sin At$ is non singular. Since the left hand side of (5.54) is a continuous function of t we get that (5.54) holds for all values of t for which $\sin At$ is non singular. We summarize these results in the following theorem, for the case when $\hbar$ and m are not necessarily equal to 1:

Theorem 5.2. *Let $\mathcal{H}$ be the separable real Hilbert space of continuous functions γ from $[0,t]$ to $\mathbb{R}^n$, such that the kinetic energy is finite with norm given by $|\gamma|^2 = \int_0^t \left(\dot{\gamma}^2 + \gamma^2\right)\,\mathrm{d}\tau$. Let B_N be the bounded symmetric operator on $\mathcal{H}$ with $D(B_N) = \mathcal{H}$ given by $(\gamma_1, B_N\gamma_2) = \langle\gamma_1, \gamma_2\rangle$, with*

$$\langle\gamma,\gamma\rangle = \frac{1}{\hbar}\int\limits_0^t \left(m\dot{\gamma}(\tau)^2 - \gamma A^2\gamma\right)\,\mathrm{d}\tau,$$

where A^2 is a strictly positive definite matrix in $\mathbb{R}^n$. Then for all values of t such that $\sin At$ is non singular we have that the range of B_N is $\mathcal{H}$ and B_N^{-1} is a bounded symmetric operator on $\mathcal{H}$. Hence, by Proposition 4.2, $D = \mathcal{H}$ and $\triangle_N$ is uniquely given by B_N. Let now φ_1, φ_2 and V be in $\mathcal{F}(\mathbb{R}^n)$ then

$$f(\gamma) = \mathrm{e}^{-\frac{i}{\hbar}\int\limits_0^t V(\gamma(\tau))\,\mathrm{d}\tau}\overline{\varphi}_1(\gamma(t))\varphi_2(\gamma(0))$$

is in $\mathcal{F}(\mathcal{H})$ and

$$\left(\varphi_1, \mathrm{e}^{-itH}\varphi_2\right) = \left|\frac{i}{2\pi}A\sin At\right|^{-\frac{1}{2}}\widetilde{\int\limits_{\mathcal{H}}}\,\mathrm{e}^{\frac{i}{2\hbar}\int\limits_0^t\left(m\dot{\gamma}^2 - \gamma A^2\gamma\right)\,\mathrm{d}\tau - \frac{i}{\hbar}\int\limits_0^t V(\gamma(\tau))\,\mathrm{d}\tau}$$

$$\overline{\varphi}_1(\gamma(t))\varphi_2(\gamma(0))\,\mathrm{d}\gamma,$$

if φ_1 and φ_2 are in $\mathcal{F}(\mathbb{R}^n) \cap L_2(\mathbb{R}^n)$.

Remark 5.1. If A^2 is not necessarily positive definite but only non degenerate as a transformation in $\mathbb{R}^n$, then with A as the unique square root of A^2 with non negative imaginary part both Theorems 5.1 and 5.2 still hold in the following sense. The Feynman path integrals are still well defined for all values of t such that $\cos At$ respectively $\sin At$ are non singular, and we may use Proposition 4.5 to prove that the operators so defined form a semigroup under t, and we can also prove, since the expansion in powers of V converges, that it is a group of unitary operators in $L_2(\mathbb{R}^n)$, by using the same method as in [30],3). Further, by computing directly the derivative with respect to t of $\left(\varphi_1, \mathrm{e}^{-itH}\varphi_2\right)$ as given in Theorem 5.2, we get that the infinitesimal generator for this unitary group is actually a self adjoint extension of

$$\left(-\triangle + xA^2x + V(x)\right)$$

defined on $\mathcal{S}(\mathbb{R}^n)$. The point of this remark is to show that the theory of Fresnel integrable functions applied to quantum mechanics via Feynman path integral has much wider applications than the more classical treatment by analytic continuation from real to imaginary time and then treating the corresponding heat equation by integration over the Wiener measure space.

Notes

This application of Feynman path integrals, first developed in the first edition of this book, concerns potentials in the class "harmonic oscillator plus bounded potential". It has given rise to further developments in connection with the trace formula for Schrödinger operators [67, 66] (see Sect. 10.3). Related work by Davies and Truman concerns relations with statistical mechanics [204]. Extensions to include polynomially growing potentials are in [105, 104], see also Sect. 10.2.

Expectations with Respect to the Ground State of the Harmonic Oscillator

We consider a harmonic oscillator with a finite number of degrees of freedom. The classical action for the time interval $[0, t]$ is given by (5.1) with $V = 0$. The corresponding action for the whole trajectory is given by

$$S_0(\gamma) = \frac{1}{2} \int_{-\infty}^{\infty} \dot{\gamma}(\tau)^2 \, d\tau - \frac{1}{2} \int_{-\infty}^{\infty} \gamma A^2 \gamma \, d\tau, \tag{6.1}$$

where $\gamma(\tau)$ and A^2 are as in (5.1) and we have set, for typographical reasons, $m = 1$. Let now $\mathcal{H}$ be the real Hilbert space of real square integrable functions on $\mathbb{R}$ with values in $\mathbb{R}^n$ and norm given by

$$|\gamma|^2 = \int_{-\infty}^{\infty} \dot{\gamma}(\tau)^2 \, d\tau + \int_{-\infty}^{\infty} \gamma(\tau)^2 \, d\tau. \tag{6.2}$$

Let B be the symmetric operator in $\mathcal{H}$ given by

$$(\gamma, B\gamma) = \int_{-\infty}^{\infty} \left(\dot{\gamma}(\tau)^2 - \gamma A^2 \gamma \right) \, d\tau \tag{6.3}$$

with domain $D(B)$ equal to the functions γ in $\mathcal{H}$ with compact support. We then have, for any $\gamma \in D(B)$, that

$$S_0(\gamma) = \frac{1}{2}(\gamma, B\gamma), \tag{6.4}$$

where $(,)$ is the inner product in $\mathcal{H}$. The Fourier transform of an element γ in $\mathcal{H}$ is given by

$$\hat{\gamma}(p) = 1/\sqrt{2\pi} \int_{-\infty}^{\infty} e^{ipt} \gamma(t) \, dt \tag{6.5}$$

and the mapping $\gamma \to \hat{\gamma}$ is then an isometry of $\mathcal{H}$ onto the real subspace of functions in $L_2\left(\left(p^2 + 1\right)\,\mathrm{d}p\right)$ satisfying

$$\overline{\hat{\gamma}(p)} = \hat{\gamma}(-p) \tag{6.6}$$

and we have, for any $\gamma \in D(B)$,

$$S_0(\gamma) = \frac{1}{2}(\gamma, B\gamma) = \int_R \overline{\hat{\gamma}(p)}\left(\frac{1}{2}p^2 - \frac{1}{2}A^2\right)\hat{\gamma}(p)\,\mathrm{d}p. \tag{6.7}$$

Moreover the range $R(B)$ of B consists of functions whose Fourier transforms are smooth functions and in $L_2\left[\left(p^2 + 1\right)\,\mathrm{d}p\right]$. Let D be the real Banach space of functions in $\mathcal{H}$ whose Fourier transforms are continuously differentiable functions with norm given by

$$\|\gamma\| = |\gamma| + \sup_p \left|\frac{\mathrm{d}\hat{\gamma}}{\mathrm{d}p}(p)\right|. \tag{6.8}$$

We have obviously that the norm in D is stronger than the norm in $\mathcal{H}$ and that D contains the range of B. We now define on $D \times D$ the symmetric form

$$\triangle(\gamma_1, \gamma_2) = \lim_{\varepsilon \to 0} \int_R \overline{\hat{\gamma}_1(p)}\left(p^2 - A^2 + i\varepsilon\right)^{-1}\hat{\gamma}_2(p)\left(p^2 + 1\right)^2\,\mathrm{d}p. \tag{6.9}$$

That this limit exists follows from the fact that $\overline{\hat{\gamma}_1(p)}\hat{\gamma}_2(p)$ is continuously differentiable and in $L_1\left[\left(p^2 + 1\right)\,\mathrm{d}p\right]$. That the form is continuous and bounded on $D \times D$ follows by standard results and (6.8). That the form is symmetric,

$$\triangle(\gamma_1, \gamma_2) = \triangle(\gamma_2, \gamma_1),$$

follows from (6.9) and (6.6). In fact the limit (6.9) has the following decomposition into its real and imaginary parts

$$\triangle(\gamma_1, \gamma_2) = P\int_R \overline{\hat{\gamma}_1(p)}\left(p^2 - A^2\right)^{-1}\hat{\gamma}_2(p)\left(p^2 + 1\right)^2\,\mathrm{d}p$$

$$-i\pi\int \overline{\hat{\gamma}_1(p)}\delta\left(p^2 - A^2\right)\hat{\gamma}_2(p)\left(p^2 + 1\right)^2\,\mathrm{d}p, \tag{6.10}$$

where the first integral is the principal value and hence real by (6.6). We see therefore that

$$\mathrm{Im}\triangle(\gamma, \gamma) \le 0. \tag{6.11}$$

Let now $\gamma_1 \in D$ and $\gamma_2 \in D(B)$, then

$$\triangle(\gamma_1, B\gamma_2) = \lim_{\varepsilon \to 0} \int_R \overline{\hat{\gamma}_1(p)}\left(p^2 - A^2 + i\varepsilon\right)^{-1}\left(p^2 - A^2\right)\hat{\gamma}_2(p)\left(p^2 + 1\right)\,\mathrm{d}p$$

$$= \int \overline{\hat{\gamma}_1(p)}\hat{\gamma}_2(p)\left(p^2 + 1\right)\,\mathrm{d}p.$$

So that

$$\triangle(\gamma_1, B\gamma_2) = (\gamma_1, \gamma_2). \tag{6.12}$$

We have now verified that $\mathcal{H}$, D, B and $\triangle$ satisfy the conditions in the Definition 4.1 for the Fresnel integral with respect to $\triangle$.

Hence for any function $f \in \mathcal{F}(D^*)$ we have that

$$\int_{\mathcal{H}}^{\triangle} e^{\frac{i}{2}(\gamma, B\gamma)} f(\gamma) \, d\gamma \tag{6.13}$$

is well defined and given by (4.12). It follows from (6.8) that γ_t, given by

$$(\gamma_t, \gamma) = \gamma(t),$$

is in $D \times \mathbb{R}^n$, since

$$\hat{\gamma}_t(p) = \sqrt{\frac{1}{2\pi}} \cdot \frac{e^{ipt}}{p^2 + 1}. \tag{6.14}$$

So that

$$f(\gamma) = e^{i \sum_{j=1}^{n} \alpha_j \cdot \gamma(t_j)} \tag{6.15}$$

is in $\mathcal{F}(D^*)$.

Hence we may compute (6.13) with $f(\gamma)$ given by (6.15) and we get

$$\int_{\mathcal{H}}^{\triangle} e^{\frac{i}{2}(\gamma, B\gamma)} f(\gamma) \, d\gamma = e^{\frac{i}{2} \sum_{j,k=1}^{n} \alpha_j \triangle(\gamma_{t_j}, \gamma_{t_k}) \alpha_k}. \tag{6.16}$$

From the definition of $\triangle$ we easily compute

$$\triangle(\gamma_s, \gamma_t) = \frac{1}{2iA} e^{-i|t-s|A}. \tag{6.17}$$

Hence we get that

$$\int_{\mathcal{H}}^{\triangle} e^{\frac{i}{2}(\gamma, B\gamma)} e^{i \sum_{j=1}^{n} \alpha_j \gamma(t_j)} \, d\gamma = e^{-\frac{1}{2} \sum_{j,k=1}^{n} \alpha_j (2A)^{-1} e^{-i|t_j - t_k|A} \alpha_k}. \tag{6.18}$$

Let now Ω_0 be the vacuum i.e. the function given by (5.21), and let us set in this section

$$H_0 = -\frac{1}{2}\triangle + \frac{1}{2}xA^2x - \frac{1}{2}\text{tr}A, \tag{6.19}$$

where we have changed the notation so that

$$H_0 \Omega_0 = 0 . \tag{6.20}$$

Let $t_1 \leq \ldots \leq t_n$, then we get from (6.18) and (5.32) that

$$
\left(\Omega_0, e^{i\alpha_1 x(t_1)} \ldots e^{i\alpha_n x(t_n)} \Omega_0 \right) = \overset{\triangle}{\underset{\mathcal{H}}{\int}} e^{\frac{1}{2}(\gamma, B\gamma)} e^{i \sum\limits_{j=1}^{n} \alpha_j \gamma(t_j)} \, d\gamma
$$

$$
= \overset{\triangle}{\underset{\mathcal{H}}{\int}} e^{i \int\limits_{-\infty}^{\infty} \left(\frac{1}{2}\dot{\gamma}^2 - \frac{1}{2}\gamma A^2 \gamma\right) \, d\tau} e^{i \sum\limits_{j=1}^{n} \alpha_j \gamma(t_j)} \, d\gamma, \tag{6.21}
$$

where[1]

$$e^{i\alpha x(t)} = e^{-itH_0} e^{i\alpha x} e^{itH_0} .$$

Theorem 6.1. *Let $\mathcal{H}$ be the real Hilbert space of real continuous and square integrable functions such that the norm given by*

$$|\gamma|^2 = \int\limits_{-\infty}^{\infty} \dot{\gamma}(\tau)^2 \, d\tau + \int\limits_{-\infty}^{\infty} \gamma(\tau)^2 \, d\tau$$

is finite. Let B be the symmetric operator with domain equal to the functions in $\mathcal{H}$ with compact support and given by

$$(\gamma, B\gamma) = 2S_0(\gamma) = \int\limits_{-\infty}^{\infty} \left(\dot{\gamma}(\tau)^2 - \gamma A^2 \gamma \right) \, d\tau,$$

and let D be the real Banach space of functions in $\mathcal{H}$ with differentiable Fourier transforms and norm given by (6.8), and let $\triangle$ be given by (6.9). Then $(\mathcal{H}, D, B, \triangle)$ satisfies the condition of Definition 4.1 for the integral normalized with respect to $\triangle$. Let f, g and V be in $\mathcal{F}(\mathbb{R}^n)$, then $f(\gamma(0))g(\gamma(t))$ and $\exp\left[-i \int_0^t V(\gamma(\tau)) \, d\tau \right]$ are all in $\mathcal{F}(D^*)$ and

$$
\left(f\Omega_0, e^{-itH} g\Omega_0 \right) = \overset{\triangle}{\underset{\mathcal{H}}{\int}} e^{iS_0(\gamma)} e^{i \int\limits_{0}^{t} V(\gamma(\tau)) \, d\tau} \overline{f}(\gamma(0)) g(\gamma(t)) \, d\gamma,
$$

where

$$H = H_0 + V$$

---

[1] This is (5.33) written in a different way.

Proof. The first part of the theorem is already proved. Let therefore f be in $\mathcal{F}(\mathbb{R}^n)$ i.e.

$$f(x) = \int e^{i\alpha x} \, d\nu(\alpha), \tag{6.22}$$

then

$$f(\gamma(0)) = \int e^{i\alpha\gamma(0)} \, d\nu(\alpha),$$

which is in $\mathcal{F}(D^*)$ by the definition of $\mathcal{F}(D^*)$, since $\gamma(0) = (\gamma_0, \gamma)$ and we already proved that $\gamma_0 \in D$. Hence also $g(\gamma(t))$ is in $\mathcal{F}(D^*)$. Now

$$\int_0^t V(\gamma(\tau)) \, d\tau = \int_0^t \int e^{i\alpha\gamma(\tau)} \, d\mu(\alpha) \, d\tau \tag{6.23}$$

is again in $\mathcal{F}(D^*)$ and therefore also $\exp\left[-i\int_0^t V(\gamma(\tau)) \, d\tau\right]$ belongs to $\mathcal{F}(D^*)$ by Proposition 4.1 (which states that $\mathcal{F}(D^*)$ is a Banach algebra). Since, also by Proposition 4.1, the Fresnel integral with respect to $\triangle$ is a continuous linear functional on this Banach algebra we have

$$\int_{\mathcal{H}}^{\triangle} e^{iS_0(\gamma)} e^{-i\int_0^t V(\gamma(\tau)) \, d\tau} \overline{f}(\gamma(0)) g(\gamma(t)) \, d\gamma$$

$$= \sum_{n=0}^{\infty} \frac{(-i)^n}{n!} \int_0^t \cdots \int_0^t \int_{\mathcal{H}}^{\triangle} e^{iS_0(\gamma)} V(\gamma(t_1)) \ldots V(\gamma(t_n)) \, d\gamma \, dt_1 \ldots \, dt_n. \tag{6.24}$$

Utilizing now (6.23), (6.21) and the perturbation expansion (5.35) the theorem is proved. $\qquad\square$

Theorem 6.2. *Let the notations be the same as in Theorem 6.1, and let $t_1 \leq \ldots \leq t_m$, then for $f_i \in \mathcal{S}(\mathbb{R}^n)$, $i = 1, \ldots m$*

$$\left(\Omega_0, f_1 e^{-i(t_2-t_1)H} f_2 e^{-i(t_3-t_2)H} f_3 \ldots e^{-i(t_m-t_{m-1})H} f_m \Omega_0\right)$$

$$= \int_{\mathcal{H}}^{\triangle} e^{iS_0(\gamma)} e^{-i\int_{t_1}^{t_m} V(\gamma(\tau)) \, d\tau} \prod_{j=1}^{m} f_j(\gamma(t_j)) \, d\gamma.$$

This theorem is proved by the series expansion of the function

$$\exp\left(-i\int_{t_1}^{t_m} V(\gamma(\tau)) \, d\tau\right)$$

and the fact that this series converges in $\mathcal{F}(D^*)$, in the same way as in the proof of Theorem 6.1.

Notes

This section, first presented in the first edition of this book, is geared towards quantum field theory (looking at nonrelativistic quantum mechanics as a "zero dimensional" quantum field theory). Formulae like (6.21) are typical of this view, see e.g. [425] for similar formulae in the "Euclidean approach" to quantum fields.

Expectations with Respect to the Gibbs State of the Harmonic Oscillator

Let $\mathcal{H}$, D, B and H_0 be as in the previous section and define the continuous symmetric form on $D \times D$ by

$$\triangle_\beta(\gamma_1, \gamma_2) = P \int_R \overline{\hat{\gamma}_1(p)} \left(p^2 - A^2\right)^{-1} \hat{\gamma}_2(p) \left(p^2 + 1\right)^2 \, \mathrm{d}p$$

$$- i\pi \int_R \overline{\hat{\gamma}_1(p)} \coth\left(\frac{\beta}{2}A\right) \cdot \delta\left(p^2 - A^2\right) \hat{\gamma}_2(p) \left(p^2 + 1\right)^2 \, \mathrm{d}p, \quad (7.1)$$

where $\beta > 0$ and $\coth\frac{\beta}{2}A = \left(1 - \mathrm{e}^{-\beta A}\right)^{-1}\left(1 + \mathrm{e}^{-\beta A}\right)$. We see that $\triangle(\gamma_1, \gamma_2) = \triangle_\infty(\gamma_1, \gamma_2)$, and we verify in the same way as in the previous section that $\mathcal{H}$, D, B and $\triangle_\infty$ satisfy the conditions of Definition 4.1 for the integral on $\mathcal{H}$ normalized with respect to $\triangle_\beta$, so that we have in particular that

$$\mathrm{Im}\,\triangle_\beta(\gamma, \gamma) \leq 0 \tag{7.2}$$

and, for $\gamma_2 \in D(B)$,

$$\triangle_\beta(\gamma_1, B\gamma_2) = (\gamma_1, \gamma_2). \tag{7.3}$$

From a direct computation we get, with γ_s defined as in the previous section (6.14), that

$$\triangle_\beta(\gamma_s, \gamma_t) = \left(2Ai\left(1 - \mathrm{e}^{-\beta A}\right)\right)^{-1}\left[\mathrm{e}^{-\mathrm{i}|t-s|A} + \mathrm{e}^{-\beta A}\mathrm{e}^{\mathrm{i}|t-s|A}\right]. \tag{7.4}$$

Set now

$$g_\beta(s - t) = \triangle_\beta(\gamma_s, \gamma_t). \tag{7.5}$$

We may then compute the Fresnel integral

$$\int_\mathcal{H}^{\triangle_\beta} \mathrm{e}^{\mathrm{i}S_0(\gamma)}\mathrm{e}^{\mathrm{i}\sum\limits_{j=1}^{m}\alpha_j\gamma(t_j)} \, \mathrm{d}\gamma = \mathrm{e}^{-\frac{\mathrm{i}}{2}\sum\limits_{j,k=1}^{m}\alpha_j g_\beta(t_j - t_k)\alpha_k}. \tag{7.6}$$

From Theorem 2.1 of [33],3) and (2.24) of the same reference we then have, for $t_1 \leq \ldots \leq t_m$,

$$
\int_{\mathcal{H}}^{\triangle_\beta} e^{iS_0(\gamma)} e^{i \sum_{j=1}^{m} \alpha_j \gamma(t_j)} \, d\gamma
$$

$$
= \left(\operatorname{tr} e^{-\beta H_0} \right)^{-1} \operatorname{tr} \left(e^{i\alpha_1 x(t_1)} \ldots e^{i\alpha_m x(t_m)} e^{-\beta H_0} \right), \tag{7.7}
$$

where $e^{i\alpha x(t)} = e^{-itH_0} e^{i\alpha x} e^{itH_0}$. And from this it follows that, for $f_i \in \mathcal{S}(\mathbb{R}^n)$, $i = 1, \ldots, n$ and with $t_1 \leq \ldots \leq t_m$,

$$
\omega_\beta^0(f_1(t_1) \ldots f_m(t_m)) = \int_{\mathcal{H}}^{\triangle_\beta} e^{iS_0(\gamma)} \prod_{j=1}^{m} f_j(\gamma(t_j)) \, d\gamma, \tag{7.8}
$$

where ω_β^0 is the Gibbs state of the harmonic oscillator i.e. for any bounded operator C on $L_2(\mathbb{R}^n)$

$$
\omega_\beta^0(C) = \left(\operatorname{tr} e^{-\beta H_0} \right)^{-1} \operatorname{tr} \left(C e^{-\beta H_0} \right). \tag{7.9}
$$

Theorem 7.1. *Let $t_1 \leq \ldots \leq t_m$ and $f_i \in \mathcal{F}(\mathbb{R}^n)$, $i = 1, \ldots, m$. If*

$$
H = H_0 + V
$$

with $V \in \mathcal{F}(\mathbb{R}^n)$ then

$$
\omega_\beta^0 \left(f_1 e^{-i(t_2 - t_1)H} f_2 \ldots e^{-i(t_m - t_{m-1})H} f_m \right)
$$

$$
= \int_{\mathcal{H}}^{\triangle_\beta} e^{iS_0(\gamma)} e^{-i \int_{t_1}^{t_m} V(\gamma(\tau)) \, d\tau} \prod_{j=1}^{m} f_j(\gamma(t_j)) \, d\gamma.
$$

Proof. This theorem is again proved by series expansion of the function $\exp\left(-i \int_{t_1}^{t_m} V(\gamma(\tau)) \, d\tau \right)$ and the fact that this series converges in $\mathcal{F}(D^*)$ in complete analogy with the proof of Theorems 6.1 and 6.2. □

Let now $0 < f(\lambda) < 1$ be a positive continuous function defined on the positive real axis, and let us define, in conformity with (7.1), the symmetric continuous form $D \times D$

$$
\triangle_f(\gamma_1, \gamma_2) = P \int_{R} \overline{\hat{\gamma}_1(p)} \left(p^2 - A^2 \right)^{-1} \hat{\gamma}_2(p) \left(p^2 + 1 \right)^2 \, dp
$$

$$
- i\pi \int_{R} \overline{\hat{\gamma}_1(p)} \frac{1 + f(A)}{1 - f(A)} \delta \left(p^2 - A^2 \right) \hat{\gamma}_2(p) \left(p^2 + 1 \right)^2 \, dp, \tag{7.10}
$$

so that $\triangle_\beta$ is equal $\triangle_f$ for $f(\lambda) = e^{-\beta\lambda}$.

It follows again easily that $\mathcal{H}$, D, B, and $\triangle_f$ satisfy the conditions of Definition 4.1 for the integral normalized with respect to $\triangle_f$, and by computation we get

$$\triangle_f(\gamma_s, \gamma_t) = (2Ai(1 - f(A)))^{-1} \left[e^{-i|t-s|A} + f(A)e^{i|t-s|A} \right]. \qquad (7.11)$$

Therefore

$$\int_{\mathcal{H}}^{\triangle_f} e^{iS_0(\gamma)} e^{i \sum\limits_{j=1}^{m} \alpha_j \gamma(t_j)} \, d\gamma = e^{-\frac{1}{2} \sum\limits_{jk} \alpha_j g_f(t_j - t_k)\alpha_k}, \qquad (7.12)$$

with $g_f(s - t) = \triangle_f(\gamma_s, \gamma_t)$.

Since the $e^{i\alpha x(t)} = e^{-itH_0} e^{i\alpha x} e^{itH_0}$ span the so called Weyl algebra on $\mathbb{R}^n$ as t and α varies, because $x(t) = \cos At \cdot x + i\frac{\sin At}{A}\pi$, where $\pi = \frac{1}{i}\frac{d}{dt}x(t)$ at $t = 0$, we may define a linear functional on the Weyl algebra by setting, for $t_1 \leq \ldots \leq t_m$

$$w_f^0 \left(e^{i\alpha_1 x(t_1)} \ldots e^{i\alpha_m x(t_m)} \right) = e^{-\frac{1}{2} \sum \alpha_j g_f(t_i - t_j)\alpha_k}. \qquad (7.13)$$

We can verify that (7.13) is consistent with the commutation relations for the Weyl algebra, and moreover w_f^0 defines a normalized positive definite state on the Weyl algebra, which is a quasi free state in the terminology of [34], 2), 3).

In fact any quasi free state invariant under the group of automorphisms induced by e^{-itH_0} is of this form, and by (7.12) we have

$$w_f^0 \left(e^{i\alpha_1 x(t_1)} \ldots e^{i\alpha_m x(t_m)} \right) = \int_{\mathcal{H}}^{\triangle_f} e^{iS_0(\gamma)} e^{i \sum\limits_{j=1}^{m} \alpha_j \gamma(t_j)} \, d\gamma. \qquad (7.14)$$

We shall return to these considerations in greater details in the next section.

Notes

This application of rigorous Feynman path integrals, first presented in the first edition of this book, connects Gibbs states of quantum statistical mechanics with time dependent observables. Corresponding formulae in an "Euclidean approach" to quantum statistical mechanics were derived before, going back to work in [33]. For recent developments see e.g.[100, 101, 98, 99]. Another connection is with Gibbs states in relativistic quantum field theory and quantum fields on curved space–times, see [33, 3] and [233].

8

The Invariant Quasi-free States

In this section we consider the harmonic oscillator with an infinite number of degrees of freedom. Hence let h be a real separable Hilbert space and A^2 be a positive self-adjoint operator on h such that zero is not an eigenvalue of A^2. The harmonic oscillator in h with harmonic potential $\frac{1}{2}x \cdot A^2 x$, where $x \cdot y$ is the inner product in h, has the classical action given by

$$S(\gamma) = \frac{1}{2} \int\limits_{-\infty}^{\infty} \dot{\gamma}(\tau)^2 \, d\tau - \frac{1}{2} \int\limits_{-\infty}^{\infty} \gamma(\tau)^2 \cdot A^2 \gamma(\tau) \, d\tau, \qquad (8.1)$$

where $\dot{\gamma}(\tau)^2 = \dot{\gamma}(\tau) \cdot \dot{\gamma}(\tau)$, and $\dot{\gamma}(\tau)$ is the strong derivative in h of the trajectory $\gamma(\tau)$, where $\gamma(\tau)$ is a continuous and differentiable function from $\mathbb{R}$ to h. The corresponding quantum mechanical system is well known and easiest described in terms of the so called annihilation-creation operators.[1] The Hamiltonian is formally given by

$$H_0 = -\frac{1}{2}\triangle + \frac{1}{2}x \cdot A^2 x - \frac{1}{2}\operatorname{tr} A, \qquad (8.2)$$

where H_0 is so normalized that

$$H_0 \Omega_0 = 0, \qquad (8.3)$$

Ω_0 being the ground state of the harmonic oscillator. The precise definition of (8.2) is in terms of annihilation-creation operators as follows.

Let $\varphi(y)$ be the self adjoint operator which is the quantization of the function $y \cdot x$ on h and $\pi(y)$ its canonical conjugate, then $\varphi(x)$ and $\pi(x)$ are given in terms of the annihilation-creation operators a^* and a by

$$\varphi(x) = a^*\left((2A)^{-\frac{1}{2}}x\right) + a\left((2A)^{-\frac{1}{2}}x\right)$$

$$\pi(x) = i\left[a^*\left(\left(\frac{1}{2}A\right)^{\frac{1}{2}}x\right) - a\left(\left(\frac{1}{2}A\right)^{\frac{1}{2}}x\right)\right] \qquad (8.4)$$

---

[1] See e.g. [53], [25],2).

for x in the domain of $A^{-\frac{1}{2}}$ and $A^{\frac{1}{2}}$ respectively. The annihilation-creation operators $a^*(x)$ and $a(x)$ are linear in x and defined for all $x \in h$, and satisfy

$$[a(x), a(y)] = [a^*(x), a^*(y)] = 0 \quad \text{and}$$
$$[a(x), a^*(y)] = x \cdot y, \tag{8.5}$$

which together with (8.4) gives

$$[\pi(x), \pi(y)] = [\varphi(x), \varphi(y)] = 0 \quad \text{and}$$
$$[\pi(x), \varphi(y)] = \frac{1}{i} x \cdot y. \tag{8.6}$$

A representation of the algebra generated by the annihilation-creation operators is provided by introducing a cyclic element Ω_0 such that

$$a(x)\Omega_0 = 0 \tag{8.7}$$

for all $x \in h$.

Let h^c be the complexification of h. We extend by linearity a^* and a to h^c and define the self adjoint operator

$$B_0(z) = a^*(z) + a(\bar{z}), \tag{8.8}$$

where $z = x + iy \in h^c$ and $\bar{z} = x - iy$ if x and y are in h. We then have the commutation relations

$$[B_0(z_1), B_0(z_2)] = i\sigma(z_1, z_2), \tag{8.9}$$

where $\sigma(z_1, z_2) = \operatorname{Im}(z_1, z_2)$, and $(z_1, z_2) = \bar{z}_1 \cdot z_2$ is the inner product in h^c. (8.7) then gives us the so called free Fock representation of the canonical commutation relations. The Weyl algebra over h^c is the *-algebra generated by elements $\epsilon(z)$ with $z \in h^c$, where the *-operator is given by $\epsilon(z)^* = \epsilon(-z)$ and the multiplication is given by

$$\epsilon(z_1) \cdot \epsilon(z_2) = e^{-i\sigma(z_1, z_2)} \epsilon(z_1 + z_2), \tag{8.10}$$

where $\sigma(z_1 \cdot z_2) = \operatorname{Im}(z_1, z_2)$ and (z_1, z_2) is the positive definite inner product in h^c.[2] It follows then from (8.9) that $\epsilon(z) \to e^{iB_0(z)}$ provides a *-representation of the Weyl algebra, which is called the free Fock representation of the Weyl algebra, and is given by the state (denoting by $\|\ \|$ the norm in h^c)

$$\omega_0(\epsilon(z)) = \left(\Omega_0, e^{iB_0(z)}\Omega_0\right) = e^{-\frac{1}{2}\|z\|^2}. \tag{8.11}$$

---

[2] See e.g. [54].

A quasi-free state on the Weyl algebra is a state given by

$$\omega_s(\epsilon(z)) = e^{-\frac{1}{2}\sigma(z,z)}, \tag{8.12}$$

where s is a real symmetric and positive definite form on $h^c = h \oplus h$ as a real Hilbert space. For a discussion of the quasi-free states see [34], 2) and [34], 3). It follows from (8.10) that $\epsilon(0) = 1$ and therefore also that $\epsilon(-z) = \epsilon(z)^{-1}$, which is equal to $\epsilon(z)^*$. Hence in any *-representation of the Weyl algebra $\epsilon(z)$ is represented by a unitary operator. Since $\omega_s(\epsilon(z)) = \omega_s(\epsilon(-z)) = \omega_s(\epsilon(z)^*)$ we have that any state of the form (8.12) gives a *-representation and therefore $\epsilon(z)$ is represented in the form $e^{iB_s(z)}$ where the $B_s(z)$ are self-adjoint and satisfy the commutation relations (8.9). It follows then that (8.12) is equivalent with

$$\omega_s(B_s(z_1)B_s(z_2)) = s(z_1, z_2) + i\sigma(z_1, z_2). \tag{8.13}$$

From the fact that

$$\omega_s([B_s(z_1) + iB_s(z_2)][B_s(z_1) - iB_s(z_2)]) \geq 0 \tag{8.14}$$

we get that

$$|\sigma(z_1, z_2)| \leq s(z_1, z_1)^{\frac{1}{2}} s(z_2, z_2)^{\frac{1}{2}}, \tag{8.15}$$

which must be satisfied in order for ω_s to be a positive state on the Weyl algebra. On the other hand, if (8.15) is satisfied, then ω_s defines a positive state on the Weyl algebra, and these states are the quasi-free states.

In this section we shall be concerned with the quasi-free states for the Weyl algebra of the harmonic oscillator which are time invariant, and for this reason we shall first define the Hamiltonian (8.2).

Since A is self adjoint we have that e^{itA} is a strongly continuous unitary group on h^c, and $\alpha_t(\epsilon(z)) = \epsilon\left(e^{itA}z\right)$ then gives a one parameter group of *-automorphisms of the Weyl algebra. Since $\omega_0(\epsilon(z)) = e^{-\frac{1}{2}\|z\|^2}$ is obviously invariant under α_t, we get that α_t induces a strongly continuous unitary group e^{itH_0} in the representation given by ω_0. Hence Ω_0 is an eigenvector with eigenvalue zero for H_0, and one finds easily that H_0 is a positive self adjoint operator. This is then the usual definition of the Hamiltonian H_0 for the harmonic oscillator, in the free Fock representation.

Since e^{itH_0} is induced by a group of *-automorphisms α_t of the Weyl algebra leaving ω_0 invariant, we may consider α_t as the group of time isomorphisms of the Weyl algebra for the harmonic oscillator. Any state of the Weyl algebra invariant under α_t will give a representation in which α_t is unitarily induced and therefore such a representation will also carry a representative for the energy of the harmonic oscillator, i.e. a Hamiltonian. We shall therefore be interested in characterizing the quasi-free states invariant under α_t.

Let us first assume that A is cyclic in h, i.e. there exists a vector $\varphi_0 \in h$ which is cyclic for A, so that $P(A)\varphi_0$ is dense in h, where $P(A)$ is an arbitrary polynomial in A. If A is not cyclic in h, then h decompose into a direct sum of closed invariant subspaces each of which is cyclic.

Since φ_0 is cyclic in h, it is also cyclic in h^c and by the spectral representation theorem we have that h^c is isomorphic with $L_2(SpA, d\nu) = L_2(d\nu)$, where $d\nu$ is the spectral measure of A given by the cyclic vector φ_0 i.e., for any continuous complex function $f(\omega)$ defined on SpA,

$$(\varphi_0, f(A)\varphi_0) = \int_{SpA} f(\omega) \, d\nu(\omega). \tag{8.16}$$

This isomorphism is given by

$$f(A)\varphi_0 \longleftrightarrow f(\omega) \tag{8.17}$$

for any $f \in C(SpA)$. By (8.16) and the fact that φ_0 is cyclic, (8.17) extends by continuity to an isomorphism between h^c and $L_2(d\nu)$. It follows now from (8.17) that, since φ_0 belongs to h and is cyclic in h, the Hilbert space h is mapped onto $L_2^\mathbb{R}(d\nu)$ i.e. the real subspace consisting of real functions. From (8.17) we get

$$Af(A)\varphi_0 \longleftrightarrow \omega f(\omega),$$

hence we may take $h = L_2^\mathbb{R}(d\nu)$, A to be the multiplication by ω on $L_2^\mathbb{R}(d\nu)$ and $h^c = L_2(d\nu)$.

A quasi-free state which is given by (8.12) is invariant under α_t if and only if the form $s(z, z)$ is invariant under the transformation $z \to e^{itA_z}$, since

$$\alpha_t(\epsilon(z)) = \epsilon \left(e^{itA_z} \right).$$

We recall that $s(z, z)$ is a symmetric (real valued positive definite) form on the real symplectic space $S = h^c$ with the symplectic structure $\sigma(z_1, z_2) = \mathrm{Im}(z_1, z_2)$, where (z_1, z_2) is the inner product in the complex Hilbert space h^c, which satisfies the condition (8.15). Since e^{itA} is unitary on h^c, it leaves σ invariant and is therefore a symplectic transformation of S and so induces a *-automorphism α_t of the Weyl algebra.

That ω_s given by (8.12), is invariant under α_t, is obviously equivalent with the positive symmetric form $s(z_1, z_2)$ defined on S being invariant under the symplectic transformation $z \to e^{itA_z}$. Let us recall that $s(z_1, z_2)$ is symmetric and bilinear only on the real space $S = h^c$, i.e. bilinear only under real linear combinations.

Let now f_1 and f_2 be Fourier transforms of real bounded signed measures μ_1 and μ_2

$$f_i(\omega) = \int e^{i\omega t} \, d\mu_i(t). \tag{8.18}$$

It follows from the spectral representation theorem that

$$f_i(A) = \int e^{itA} \, d\mu_i(t) \tag{8.19}$$

Hence

$$s(f_1(A)\varphi_0, f_2(A)\varphi_0) = \int\int s\left(e^{it_1 A}\varphi_0, e^{it_2 A}\varphi_0,\right) \, d\mu_1(t_1) \, d\mu_2(t).$$

By the invariance of s under $z \to e^{itA_z}$ we then get

$$s(f_1(A)\varphi_0, f_2(A)\varphi_0) = \int\int s\left(\varphi_0, e^{i(t_2-t_1)A}\varphi_0\right) \, d\mu_1(t_1) \, d\mu_2(t_2)$$

$$= \int\int s\left(\varphi_0, e^{itA}\varphi_0\right) \, d\mu_1(t_1) \, d\mu_2(t+t_1). \tag{8.20}$$

Now

$$\overline{f_1}f_2(\omega) = \overline{f_1}(\omega) \cdot f_2(\omega) = \int\int e^{i\omega(t_2-t_1)} \, d\mu_1(t_1) \, d\mu_2(t_2)$$

$$= \int\int e^{i\omega t} \, d\mu_1(t_1) \, d\mu_2(t+t_1), \tag{8.21}$$

from which we get that

$$s\left(\varphi_0, \overline{f_1}f_2(A)\varphi_0\right) = \int\int s\left(\varphi_0, e^{itA}\varphi_0\right) \, d\mu_1(t_1) \, d\mu_2(t+t_1).$$

Hence we have proved

$$s(f_1(A)\varphi_0, f_2(A)\varphi_0) = s(\varphi_0, \overline{f_1}f_2(A)\varphi_0) \tag{8.22}$$

for any f_1 and f_2 which are Fourier transforms of bounded signed real measures on $\mathbb{R}$. Hence (8.22) holds for all f_1 and f_2 in $\mathcal{S}(\mathbb{R})$ such that $f_i(\omega) = \overline{f_i}(-\omega)$. Now we obviously have that the functions f in $\mathcal{S}(\mathbb{R})$ satisfying $f(\omega) = f(-\omega)$ are dense in $C_0[0, \infty]$, the space of continuous functions on $[0, \infty]$ tending to zero at infinity. Hence by continuity, since $Sp(A) \subset [0, \infty]$, (8.22) holds for all continuous bounded functions tending to zero at infinity. The strong continuity of $s(z, z)$ follows from the fact that $s(z_1, z_2)$ is bilinear and $s(z, z)$ is positive and defined for all z, and this gives that $s(f_1(A)\varphi_0, f_2(A)\varphi_0)$ is continuous in the strong $L_2(d\nu)$ topology. We have thus that

$$s(\varphi_0, \overline{f_1}f_2(A)\varphi_0) = s(f_1(A)\varphi_0, f_2(A)\varphi_0), \tag{8.23}$$

for continuous f_1 and f_2 being zero at infinity, and in fact by the strong $L_2(d\nu)$ continuity also for all f_1 and f_2 in $L_2(d\nu)$. From this we get that, if $g \geq 0$, then

$$s(\varphi_0, g(A)\varphi_0) = s\left(g^{\frac{1}{2}}(A)\varphi_0, g^{\frac{1}{2}}(A)\varphi_0\right), \tag{8.24}$$

so that $s(\varphi_0, g(A)\varphi_0)$ defines a bounded positive linear functional on the space of continuous functions, hence a measure which is obviously absolutely continuous with respect to the spectral measure. So we have proved that

$$s(f_1(A)\varphi_0, f_2(A)\varphi_0) = \int \overline{f_1} f_2(\omega) \rho(\omega) \ d\nu(\omega), \tag{8.25}$$

where ρ is a positive measurable function, and the right hand side is also the representation of s in the spectral representation of h^c. The condition (8.15) for the positivity of the state ω_s is obviously equivalent to the condition

$$\rho(\omega) \geq 1 \quad \text{a.e.}$$

We should remark that we only assumed $s(z_1, z_2)$ to be bilinear under real linear combinations, but in fact the invariance of s under $z \to e^{itA} z$ gives that s is of the form (8.25), which is actually a sesquilinear form. By the fact that $s(z_1, z_2)$ is everywhere defined, we get that $\rho(\omega)$ is bounded almost everywhere. So in fact we may write (8.25) also as

$$s(z_1, z_2) = (z_1, Bz_2),$$

where B is a bounded symmetric operator commuting with A, such that $B \geq 1$.

Theorem 8.1. *Let h be a real separable Hilbert space and A be a positive self adjoint operator on h such that zero is not an eigenvalue of A. Let $S = h^c$ be the real symplectic space with symplectic structure given by $\sigma(z_1, z_2) = \text{Im}(z_1, z_2)$, where (z_1, z_2) is the inner product on the complex Hilbert space h^c. $z \to e^{itA} z$ is then a group of symplectic transformations on S and generates therefore a group α_t of *-automorphisms of the Weyl algebra over S, where the Weyl algebra is the algebra generated by $\epsilon(z)$, $z \in S$ with the multiplication*

$$\epsilon(z_1)\epsilon(z_2) = e^{-i\sigma(z_1,z_2)} \epsilon(z_1 + z_2)$$

*and *-operation given by $\epsilon(z)^* = \epsilon(-z)$. A quasi-free state of the Weyl algebra is a state of the form*

$$\omega_s(\epsilon(z)) = e^{-\frac{1}{2}s(z,z)},$$

where $s(z, z)$ is a positive bilinear form on the real space S.

A necessary and sufficient condition for ω_s to be a quasi-free state invariant under α_t, is that there exists a bounded, symmetric operator C on the complex Hilbert space h^c such that $C \geq 1$, C commutes with A and

$$s(z_1, z_2) = (z_1, Cz_2),$$

where $(,)$ is the inner product in the complex Hilbert space h^c.

Proof. If A is cyclic in h then the theorem is already proved. If A is not cyclic in h, h decomposes in a direct sum $h = \bigoplus_i h_i$, and, in each component h_i, A is cyclic, i runs over at most a countable index set since h is separable. Let φ_i be a cyclic vector in h_i. Then, with f_i continuous, $\sum_{i=1}^n f_i(A)\varphi_i$ is dense in h, and in the same way as in the cyclic case we prove that

$$s\left(\sum_{i=1}^n f_i(A)\varphi_i, \sum_i g_i(A)\varphi_i\right) = \sum_{ij} s\left(\varphi_i, \overline{f_i}(A)f_j(A)\varphi_j\right)$$

$$= \int \overline{f_i}(\omega)\rho_{ij}(\omega)f_j(\omega)\ \mathrm{d}\nu(\omega).$$

The last line is actually the spectral resolution $\rho_{ij}(\omega)$ of a operator C that commutes with A. This proves the theorem.   □

For simplicity of notation we shall now assume that A acts cyclic in h. This is in reality no restriction since, if not, then h decomposes, $h = \bigoplus_n h_n$, in at most a countable sum of cyclic subspaces.

Let now $\mathcal{H}$ be the real Hilbert space of h valued functions on $\mathbb{R}$ which are continuous and such that the norm $|\gamma|$ is finite, where

$$|\gamma|^2 = \int\limits_{-\infty}^{\infty} \left(\dot{\gamma}\cdot\dot{\gamma} + \gamma A^2\gamma\right)\ \mathrm{d}\tau$$

and $\gamma\cdot\gamma$ is the inner product in h. On this Hilbert space the classical action $S(\gamma)$ for the harmonic oscillator

$$S(\gamma) = \frac{1}{2}\int\limits_{-\infty}^{\infty} \dot{\gamma}\cdot\dot{\gamma}\ \mathrm{d}\tau - \frac{1}{2}\int\limits_{-\infty}^{\infty} \gamma(\tau)\cdot A^2\gamma(\tau)\ \mathrm{d}\tau$$

is a bounded quadratic form. Let φ be a cyclic vector for A in h, we know then that h may be identified with $L_2^{\mathbb{R}}(\mathrm{d}\nu)$, and therefore $\mathcal{H}$ with the real functions in two real variables with norm

$$|\gamma|^2 = \int\int \left(\left(\frac{\partial\gamma}{\partial t}\right)^2 + \omega^2\gamma^2(t,\omega)\right)\ \mathrm{d}t\ \mathrm{d}\nu(\omega), \tag{8.28}$$

and recalling that zero is not an eigenvalue of A, so that the set $\{0\}$ has ν-measure zero, we see that (8.28) defines a Hilbert norm. Introducing now the Fourier transform $\hat{\gamma}(p,\omega)$ of $\gamma(t,\omega)$ with respect to t, we get

$$|\gamma|^2 = \int\int |\hat{\gamma}(p,\omega)|^2(p^2+\omega^2)\ \mathrm{d}p\ \mathrm{d}\nu(\omega), \tag{8.29}$$

so that $\gamma \to \hat{\gamma}$ is an isometry of $\mathcal{H}$ onto the real subspace of $L_2^{\mathbb{C}}[(p^2 + \omega^2)\ \mathrm{d}p\ \mathrm{d}\nu(\omega)]$ consisting of functions satisfying

$$\overline{\hat{\gamma}(p,\omega)} = \hat{\gamma}(-p,\omega). \tag{8.30}$$

Let now D be the subspace of functions in $\mathcal{H}$ consisting of functions γ such that $\hat{\gamma}(p, \omega)$ is continuous in ω and continuously differentiable in p with norm

$$\|\gamma\| = |\gamma| + \sup_{\omega, p} \left| \frac{\partial \hat{\gamma}}{\partial p}(p, \omega) \right|. \tag{8.31}$$

We define a bounded symmetric operator B in $\mathcal{H}$ by

$$(\gamma, B\gamma) = 2S(\gamma), \tag{8.32}$$

with domain $D(B)$ consisting of functions $\gamma(t, \omega)$ which are continuous and with compact support in R^2. For $\omega \in D(B)$ we have that

$$\hat{\gamma}(p, \omega) = \frac{1}{\sqrt{2\pi}} \int e^{ipt} \gamma(t, \omega) \, dt \tag{8.33}$$

is obviously continuous in ω and p and continuously differentiable in p. Now the Fourier transform of $B\gamma$ is, by (8.32), given by

$$\widehat{B\gamma}(p, \omega) = \frac{p^2 - \omega^2}{p^2 + \omega^2} \, \hat{\gamma}(p, \omega). \tag{8.34}$$

From this it follows that the range $R(B)$ of B consists of functions, the Fourier transform of which are continuously differentiable in p with uniformly bounded derivatives and continuous in ω. Hence we have

$$R(B) \subset D. \tag{8.35}$$

Let now $C \geq 0$ be a bounded symmetric operator on h commuting with A. Since A acts cyclically in h, we have that C is represented by a bounded measurable function $c(\omega) \geq 0$ a.e. We now define the symmetric and continuous form $\triangle_C(\gamma_1, \gamma_2)$ on $D \times D$ by

$$\triangle_C(\gamma_1, \gamma_2) = \int d\nu(\omega) \left\{ P \int_{\mathbb{R}} \overline{\hat{\gamma}_1(p, \omega)} \frac{(p^2 + \omega^2)^2}{p^2 - \omega^2} \hat{\gamma}_2(p, \omega) \, dp \right.$$

$$\left. -i\pi c(\omega) \int_{\mathbb{R}} \overline{\hat{\gamma}_1(p, \omega)} \delta(p^2 - \omega^2)(p^2 + \omega^2)^2 \hat{\gamma}_2(p, \omega) \, dp \right\}. \tag{8.36}$$

From (8.34) we have, for $\gamma_1 \in D$ and $\gamma_2 \in D(B)$,

$$\triangle_C(\gamma_1, B\gamma_2) = (\gamma_1, \gamma_2). \tag{8.37}$$

Since $c(\omega) \geq 0$ we also have that $\triangle_C$ has non positive imaginary part, so that $\mathcal{H}$, D, B and $\triangle_C$ satisfy the conditions of Definition 4.1 for the integral on $\mathcal{H}$ normalized with respect to $\triangle_C$.

Let now $u \in h$, we define the element $\gamma_u^s(t) \in \mathcal{H}$ by

$$(\gamma_u^s, \gamma) = u \cdot \gamma(s) \tag{8.38}$$

and we have then that

$$\gamma_u^s(t) = \frac{1}{2A} e^{-|t-s|A} \cdot u, \tag{8.39}$$

so that

$$(\gamma_u^s, \gamma_u^s) = \frac{1}{2} u \cdot A^{-1} u, \tag{8.40}$$

which implies that $\gamma_u^s \in \mathcal{H}$ if $u \in D\left(A^{-\frac{1}{2}}\right)$. Furthermore we get by computation that

$$\hat{\gamma}_u^s(p, \omega) = \frac{1}{\sqrt{2\pi}} \frac{e^{ips}}{p^2 + \omega^2} \cdot u(\omega), \tag{8.41}$$

hence, for $u(\omega)$ continuous and bounded and in $D\left(A^{-\frac{1}{2}}\right)$, that $\hat{\gamma}_u^s(p, \omega)$ is in D. Moreover further computations give

$$\triangle_C(\gamma_u^s, \gamma_v^t) = -u \cdot \left[\frac{1}{2A} \sin|t - s|A + \frac{i}{2A} C \cos(t - s)A \right] v. \tag{8.42}$$

Let now $G_C(s - t)$ be the self adjoint operator in h defined by

$$G_C(s - t) = -\frac{1}{2A} \left[\sin|t - s|A + iC \cos(t - s)A \right]. \tag{8.43}$$

Then

$$\triangle_C(\gamma_u^s, \gamma_v^t) = u \cdot G_C(s - t)v. \tag{8.44}$$

Let $u_1, \ldots, u_n$ be in $D\left(A^{-\frac{1}{2}}\right)$ and such that $u_1(\omega), \ldots, u_n(\omega)$ are continuous and bounded. Then $\gamma_{u_i}^{t_i} \in D$ for $i = 1, \ldots, n$. Hence, with

$$f(\gamma) = e^{i \sum\limits_{j=1}^{n} u_j \cdot \gamma(t_j)} = e^{i \sum\limits_{j=1}^{n} \left(\gamma, \gamma_{u_j}^{t_j}\right)},$$

we get $f(\gamma) \in \mathcal{F}(D^*)$, so that we may compute

$$\int\limits_{\mathcal{H}}^{\triangle_C} e^{iS(\gamma)} e^{i \sum\limits_{j=1}^{n} u_j \cdot \gamma(t_j)} \, d\gamma = e^{-\frac{1}{2} \sum\limits_{jk} u_j G_C(t_j - t_k) u_k}. \tag{8.45}$$

Let us now consider the quasi-free state of Theorem 8.1,

$$\omega_C(\epsilon(z)) = e^{-\frac{1}{2}(z, Cz)}, \tag{8.46}$$

where $C \geq 1$ and commutes with A. By (8.4) we have that, for $u \in D\left(A^{-\frac{1}{2}}\right)$,

$$\varphi(u) = a^*\left((2A)^{-\frac{1}{2}}u\right) + a\left((2A)^{-\frac{1}{2}}u\right) \tag{8.47}$$

is the quantization of the linear function $u \cdot x$ defined on h. In conformity with the notation used in Chap. 7 we define

$$u \cdot x(t) = \alpha_t(\varphi(u)), \tag{8.48}$$

where α_t is the group of time automorphisms given by (8.20). In fact we have then that, expressed in the Weyl algebra for the harmonic oscillator,

$$\exp(iu \cdot x(t)) = \epsilon\left(e^{itA}(2A)^{-\frac{1}{2}}u\right). \tag{8.49}$$

Let now $u_1, \ldots, u_n$ be in $D\left(A^{-\frac{1}{2}}\right)$ and consider, for $t_1 \leq \ldots \leq t_n$,

$$\omega_C\left(e^{iu_1 \cdot x(t_1)} \ldots e^{iu_n \cdot x(t_n)}\right). \tag{8.50}$$

We get easily, using (8.10) and (8.46), that (8.50) is equal to

$$\omega_C\left(e^{iu_1 \cdot x(t_1)} \ldots e^{iu_n \cdot x(t_n)}\right) = e^{-\frac{i}{2}\sum_{jk} u_j \cdot G_C(t_j - t_k)u_k}. \tag{8.51}$$

So by (8.45) we have proved

$$\omega_C\left(e^{iu_1 \cdot x(t_1)} \ldots e^{iu_n \cdot x(t_n)}\right) = \int_{\mathcal{H}}^{\Delta_C} e^{iS(\gamma)} e^{i\sum_{j=1}^{n} u_j \cdot \gamma(t_j)} \, d\gamma. \tag{8.52}$$

We state this fact in the following theorem:

Theorem 8.2. *Let h be a real separable Hilbert space and A a positive self adjoint operator on h such that zero is not an eigenvalue of A. The classical action for the harmonic oscillator on h is given by*

$$S(\gamma) = \frac{1}{2} \int_{-\infty}^{\infty} \left(\dot{\gamma} \cdot \dot{\gamma} - \gamma \cdot A^2 \gamma\right) \, dt.$$

Let α_t be the time automorphism of the Weyl algebra for the corresponding quantum system. Let $C \geq 0$ be a bounded self adjoint operator commuting with A, then the Fresnel integral relative to $2S(\gamma)$, normalized with respect to Δ_C, *where $\triangle_C$ is given in (8.36), exists and for $C \geq 1$ this Fresnel integral induces a quasi-free state on the Weyl algebra, invariant under α_t, by the formula*

$$\omega_C\left(e^{iu_1 \cdot x(t_1)} \dots e^{iu_n \cdot x(t_n)}\right) = \int_{\mathcal{H}}^{\triangle_C} e^{iS(\gamma)} e^{i \sum_{j=1}^{n} u_j \gamma(t_j)} \, d\gamma,$$

with $u_1, \dots, u_n$ in $D\left(A^{-\frac{1}{2}}\right)$ and $t_1 \leq \dots \leq t_n$.

Moreover any invariant quasi-free state on the Weyl algebra is obtained in this way. In particular, if $C = 1$ we get the free Fock state, and if $C = \coth\left(\frac{\beta}{2}A\right)$ we get the free Gibbs-state at temperature $1/\beta$.

Proof. The first part is already proved. The moreover part follows from Theorem 8.1. That we get the Fock state for $C = 1$ follows by direct inspection and that we get the free Gibbs-state with $C = \coth\left(\frac{\beta}{2}A\right)$ follows from the form of $G_C(s-t)$ given by (8.43) and [33], 3) (3.32). This proves the theorem. $\square$

Remark. We have thus, in particular, that the Fresnel integrals relative to the quadratic form $2S(\gamma)$ on $\mathcal{H}$ correspond, in the sense of Theorem 8.2, to the linear functionals on the Weyl algebra given by

$$\omega_C(\epsilon(z)) = e^{-\frac{1}{2}(z, Cz)} \tag{8.53}$$

where $C \geq 0$ and commutes with A. However these functionals are positive states on the Weyl algebra only in the case $C \geq 1$.

Notes

These results appeared first in the first edition of this book. They have relations with the theory of quantum fields and the theory of representations of the Weyl algebra. This work has lead to new developments in these directions, also in connection with the approach of Feynman path integrals via Poisson processes, see [63] and references therein. See also the final chapter of the present book.

The Feynman History Integral
for the Relativistic Quantum Boson Field

The free relativistic scalar boson field in n space dimensions is a harmonic oscillator in the sense of the previous section, with $h = L_2^{\mathbb{R}}(\mathbb{R}^n)$ and $A^2 = -\triangle + m^2$, where $\triangle$ is the Laplacian as a self adjoint operator on $L_2^{\mathbb{R}}(\mathbb{R}^n)$ and m is a non-negative constant called the mass of the field. Because of the importance of this physical system we shall give it a more detailed treatment.

We shall first discuss the free relativistic boson field i.e. the system with a classical action given by

$$S(\varphi) = \frac{1}{2} \int \int_{\mathbb{R}^n} \left[\left(\frac{\partial \varphi}{\partial t} \right)^2 - \sum_{i=1}^n \left(\frac{\partial \varphi}{\partial x_i} \right)^2 - m^2 \varphi^2 \right] \, d\vec{x} \, dt. \tag{9.1}$$

Let $\mathcal{H} = \mathcal{H}_1$ be a real Sobolev space, namely the Hilbert space of real valued functions φ over $\mathbb{R}^{n+1}$ for which the norm $|\varphi|$ is finite:

$$|\varphi|^2 = \int_{\mathbb{R}^{n+1}} \left[\left(\frac{\partial \varphi}{\partial t} \right)^2 + \sum_{i=1}^n \left(\frac{\partial \varphi}{\partial x_i} \right)^2 + \varphi^2 \right] \, d\vec{x} \, dt. \tag{9.2}$$

Then $S(\varphi)$ is a bounded continuous quadratic form on $\mathcal{H}$ and we define a bounded symmetric operator B on $\mathcal{H}$ by

$$(\varphi, B\varphi) = 2S(\varphi), \tag{9.3}$$

with domain $D(B)$ equal to the set of functions in $\mathcal{H}$ with compact support in $\mathbb{R}^{n+1}$. The Fourier transformation $\varphi \to \hat{\varphi}$

$$\hat{\varphi}(p) = (2\pi)^{-\frac{n+1}{2}} \int_{\mathbb{R}^{n+1}} e^{ipx} \varphi(x) \, dx, \tag{9.4}$$

with $x = \{t, \vec{x}\}$, is an isomorphism of $\mathcal{H}$ with the real subspace of L_2 $((p^2 + 1) \, dp)$ consisting of functions $\hat{\varphi}$ such that

$$\overline{\hat{\varphi}}(p) = \hat{\varphi}(-p). \tag{9.5}$$

Let D be the linear subspace of functions φ in $\mathcal{H}$ whose Fourier transforms $\hat{\varphi}(p)$ are continuously differentiable with bounded derivatives. The norm in D is given by

$$\|\varphi\| = |\varphi| + \sup_{i,p} \left| \frac{\partial \hat{\varphi}}{\partial p_i} \right|. \tag{9.6}$$

If $\varphi \in D(B)$ then $\hat{\varphi}$ is a smooth function and the Fourier transform of $B\varphi$ is given by

$$\widehat{B\varphi}(p) = \frac{p_0 - \vec{p}^2 - m^2}{p^2 + 1} \hat{\varphi}(p), \tag{9.7}$$

width $p = \{p_0, \vec{p}\}$.

Now, since $\varphi \in D(B)$, we have that

$$\frac{\partial \hat{\varphi}}{\partial p_j} = (2\pi)^{-\frac{n+1}{2}} \int_C e^{ipx}(ix_j)\varphi(x)\,dx, \tag{9.8}$$

where C is compact. Since φ is in $L_2(\mathbb{R}^{n+1})$ and $C = \operatorname{supp}\varphi$ is compact, φ is also in L_1, so that $\frac{\partial \hat{\varphi}}{\partial p_j}$ is bounded and continuous. This gives immediately, from (9.7), that $B\varphi$ is in D. Let $c(\vec{p})$ be a measurable non negative function on $\mathbb{R}^n$. We then define a continuous and bounded symmetric bilinear form $\triangle_C(\varphi, \psi)$ on $D \times D$ by

$$\triangle_C(\varphi, \psi) = P \int_{\mathbb{R}^{n+1}} \overline{\hat{\varphi}}(p) \frac{(p^2 + 1)^2}{p_0^2 - \vec{p}^2 - m^2} \hat{\psi}(p)\,dp$$

$$- i\pi \int c(\vec{p})\overline{\hat{\varphi}}(p)\delta\left(p_0^2 - \vec{p}^2 - m^2\right)(p^2 + 1)^2 \hat{\psi}(p)\,dp, \tag{9.9}$$

where $P\int \frac{f(p)}{p_0^2 - \vec{p}^2 - m^2}\,dp = \int_{\mathbb{R}^n} \left[P \int_{\mathbb{R}} \frac{f(p,\vec{p})}{p_0^2 - \vec{p}^2 - m^2}\,dp_0 \right] d\vec{p}$ for any smooth function $f(p)$ and $P \int_{\mathbb{R}} \frac{f(p_0,\vec{p})}{p_0^2 - \vec{p}^2 - m^2}\,dp_0$ is the principal value integral. Since the first term in (9.9) is real, we see that

$$\operatorname{Im}\triangle_C(\varphi, \varphi) \leq 0.$$

Let now $\varphi \in D$ and $\psi \in D(B)$. Then we get from (9.9) that

$$\triangle_C(\varphi, B\psi) = (\varphi, \psi), \quad ((,): \text{scalar product in } \mathcal{H}) \tag{9.10}$$

and hence we have verified that $\mathcal{H}$, D, B and $\triangle_C$ satisfy the conditions of Definition 4.1 for the Fresnel integral with respect to the classical action $S(\varphi)$ and normalized according to $\triangle_C$. This Fresnel integral will also be called Feynman history integral, and if we want to emphasize the dependence on the non negative function c we shall call it the Feynman history integral relative to C.

Let now h^c be the complexification of $h = L_2^{\mathbb{R}}(\mathbb{R}^n)$, and consider the Weyl algebra over h^c. The quantized field $\Phi(\vec{x})$ at time zero is then given in terms of the Weyl algebra. In fact, for any $f \in h$ such that $f \in D\left(A^{-\frac{1}{2}}\right)$, we have

$$e^{i\Phi(f)} = \epsilon\left((2A)^{-\frac{1}{2}}f\right), \tag{9.11}$$

where $\Phi(f) = \int \Phi(\vec{x})f(\vec{x})\,\mathrm{d}\vec{x}$ and $A = \sqrt{-\triangle + m^2}$. Thus $\Phi(f) = \int \Phi(\vec{x})f(\vec{x})\,\mathrm{d}\vec{x}$ is understood in the operator valued distributional sense, $\Phi(f)$ being a well defined linear operator, depending on the test function f. For the definition of the Weyl algebra over h^c see the previous section. The time automorphism of the Weyl algebra was given by

$$\alpha_t^0(\epsilon(g)) = \epsilon\left(e^{itA}g\right). \tag{9.12}$$

Now $h^c = L_2(\mathbb{R}^n)$ carries a natural unitary representation of the translation group $\mathbb{R}^n$, so that, for any $a \in \mathbb{R}^n$, $g \to g_a$ with $g_a(x) = g(x - a)$ is a unitary transformation of h^c. Since it is unitary it is also symplectic, hence

$$\beta_a(\epsilon(g)) = \epsilon(g_a) \tag{9.13}$$

is a *-automorphism of the Weyl algebra. We have the following theorem.

Theorem 9.1. *Let $h = L_2^{\mathbb{R}}\mathbb{R}^n)$ and $h^c = L_2(\mathbb{R}^n)$. Let $g \in h^c$ and $\epsilon(g)$ the corresponding element in the Weyl algebra over h^c. The quantized time zero field $\Phi(f)$ is then expressed in terms of this Weyl algebra by*

$$e^{i\Phi(f)} = \epsilon\left((2A)^{-\frac{1}{2}}f\right)$$

for any $f \in h$.

Any quasi-free state which is invariant under the time automorphisms α_t^0 and also under the space automorphisms β_a is of the form

$$\omega_C(\epsilon(g)) = e^{-\frac{1}{2}(g,C_g)},$$

where

$$(g, Cg) = \int_{\mathbb{R}^n} c(\vec{p})|\hat{g}(\vec{p})|^2 \,\mathrm{d}\vec{p}$$

and $c(\vec{p})$ is a bounded measurable function such that

$$c(\vec{p}) \geq 1.$$

Proof. That $\omega_C(e(g)) = e^{-\frac{1}{2}(g,C_g)}$, where C is a bounded symmetric operator on h^c such that $C \geq 1$ and C commutes with A, follows from Theorem 8.1. Now, since ω_C is to be invariant under β_a, we get

$$\omega_C(\epsilon(g)) = \omega_C(\beta_a(\epsilon(g))) = \omega_C(\epsilon(g_a)),$$

hence

$$(g, Cg) = (g_a, Cg_a),$$

so that C is an operator in $L_2(\mathbb{R}^n)$ which commutes with translation. Hence C is of the form given in the theorem. This proves the theorem. $\qquad\square$

Let now

$$e^{i\Phi_t(f)} = \alpha_t^0 \left(e^{i\Phi(f)} \right). \qquad (9.14)$$

We have the following:

Theorem 9.2. *Let $h = L_2^{\mathbb{R}}(\mathbb{R}^n)$ and $h^c = L_2(\mathbb{R}^n)$. The classical action for the free relativistic scalar boson field in $\mathbb{R}^n$ is given by*

$$S(\varphi) = \frac{1}{2} \int\limits_{\mathbb{R}} \int\limits_{\mathbb{R}^n} \left(\left(\frac{\partial\varphi}{\partial t} \right)^2 - \sum_{j=1}^{n} \left(\frac{\partial\varphi}{\partial x_j} \right)^2 - m^2\varphi^2 \right) \, d\vec{x} \, dt \, ,$$

where m is a non negative constant called the mass of the free field. Let α_t and β_a be the time and space automorphisms of the Weyl algebra over h^c. Let $c(\vec{p}) \geq 0$ be a non negative bounded measurable function on $\mathbb{R}^n$. Then the Fresnel integral relative to $2S(\varphi)$, normalized with respect to $\triangle_C$ exists, where $\triangle_C$ is given in (9.9), and is called the Feynman history integral relative to C. If $c(\vec{p}) \geq 1$, the corresponding Feynman history integral defines a quasi-free state ω_C on the Weyl algebra for the scalar field, i.e. the Weyl algebra over h^c, and ω_C is invariant under the time and space automorphisms α_t and β_a of the Weyl algebra. The correspondence between the history integral and the state is given, for $t_1 \leq \ldots \leq t_n$ by the formula

$$\omega_C \left(e^{i\Phi_{t_1}(f_1)} \ldots e^{i\Phi_{t_n}(f_n)} \right) = \int\limits_{\mathcal{H}}^{\triangle_C} e^{iS(\varphi)} e^{i \sum\limits_{j=1}^{n} \int \varphi(\vec{x},t_j) f_j(\vec{x}) \, d\vec{x}} \, d\varphi,$$

if $f_1, \ldots, f_n$ are in $D\left(A^{-\frac{1}{2}} \right)$.

Moreover any quasi-free state invariant under space–time translations is obtained in this way. In particular, for $c(\vec{p}) = 1$ we get the free Fock representation[1] and for $c(\vec{p}) = \coth\left(\frac{\beta}{2}\omega(\vec{p}) \right)$, with $\omega(\vec{p}) = \sqrt{\vec{p}^2 + m^2}$, we get the free Gibbs state at temperature $1/\beta$.

Proof. That the Fresnel integral relative to $2S(\varphi)$ normalized with respect to $\triangle_C$ exists, with $\triangle_C$ given by (9.9), was proven before. Consider now,

---

[1] In this case $\triangle_C$ is the Feynman i.e. the causal propagator. For a discussion of its role in relativistic local quantum field theory see [55].

for $t_1 \leq t_2 \leq \ldots \leq t_n$ and $f_1, \ldots, f_n$ in $D\left(A^{-\frac{1}{2}}\right)$, $w_C\left(e^{i\Phi_{t_1}(f_1)} \ldots e^{i\Phi_{t_n}(f_n)}\right)$, where w_C is the invariant quasi-free state of Theorem 9.1 and $e^{i\Phi_t(f)}$ is defined by (9.14). We have then, using (9.14) and (9.11):

$$w_C\left(e^{i\Phi_{t_1}(f_1)} \ldots e^{i\Phi_{t_n}(f_n)}\right) = w_C\left(\alpha_{t_1}^0\left(e^{i\Phi(f_1)}\right) \ldots \alpha_{t_n}^0\left(e^{i\Phi(f_n)}\right)\right)$$
$$= w_C\left(\alpha_{t_1}^0\left(\epsilon\left((2A)^{-\frac{1}{2}}f_1\right)\right) \ldots \alpha_{t_n}^0\left(\epsilon\left((2A)^{-\frac{1}{2}}f_n\right)\right)\right).$$

Hence, using (9.12):

$$w_C\left(e^{i\Phi_{t_1}(f_1)} \ldots e^{i\Phi_{t_n}(f_n)}\right) = w_C\left(\epsilon\left(e^{it_1 A}(2A)^{-\frac{1}{2}}f_1\right) \ldots \epsilon\left(e^{it_n A}(2A)^{-\frac{1}{2}}f_n\right)\right),$$

and therefore, from the property (8.10) of the multiplication in the Weyl algebra and the fact that, by Theorem 9.1,

$$w_C(\epsilon(g)) = e^{-\frac{1}{2}(g, Cg)}.$$

we have:

$$w_C\left(e^{i\Phi_{t_1}(f_1)} \ldots e^{i\Phi_{t_n}(f_n)}\right) = e^{-\frac{1}{2}\sum_{jk}\int \overline{\hat{f}_j(\vec{p})}\hat{G}_C(t_j - t_k, \vec{p})\hat{f}_k(\vec{p})\ d\vec{p}} \tag{9.15}$$

where

$$\hat{G}_C(t, \vec{p}) = \frac{-1}{2\sqrt{\vec{p}^2 + m^2}}\left(\sin|t|\sqrt{\vec{p}^2 + m^2} + ic(\vec{p})\cos|t|\sqrt{\vec{p}^2 + m^2}\right). \tag{9.16}$$

On the other hand we get easily that

$$F(\varphi) = e^{i\sum_{j=1}^{n}\int \varphi(\vec{x}, t_j)f_j(\vec{x})\ d\vec{x}} \tag{9.17}$$

is in $\mathcal{F}(D^*)$ so that we may compute

$$\int_{\mathcal{H}}^{\triangle_C} e^{iS(\varphi)} e^{i\sum_{j=1}^{n}\int \varphi(\vec{x}, t_j)f_j(\vec{x})\ d\vec{x}}\ d\varphi \tag{9.18}$$

Using now (9.9) for $\triangle_C$ and a representation of the form (8.39) for the linear functional $\int \varphi(\vec{x}, t)f(\vec{x})\ d\vec{x}$ defined on $\mathcal{H}$, we get, with

$$(\psi_t(f), \varphi) \equiv \int \varphi(\vec{x}, t)f(\vec{x})\ d\vec{x}, \tag{9.19}$$

that $\psi_t(f)$ is in D, so that $F(\varphi)$ is in $\mathcal{F}(D^*)$ and

$$\triangle_C(\psi_s(f), \psi_t(g)) = \int \hat{G}_C(s - t, \vec{p})\overline{\hat{f}(\vec{p})}\hat{g}(\vec{p})\ d\vec{p}. \tag{9.20}$$

Hence we obtain from (9.20) and (9.15) the identity in the theorem. That any space and time invariant quasi free state is obtained in this way, follows from the previous theorem. That we get the free vacuum or free Fock representation for $c(\vec{p}) = 1$ is standard and that we get the free Gibbs state at temperature $1/\beta$ if $c(\vec{p}) = \coth\left(\frac{\beta}{2}\omega(\vec{p})\right)$ is proved in [33], 3) Chap. 3 (3.36). This then proves the theorem.   □

Let now ψ be a smooth non negative function on $\mathbb{R}^n$ so that $\int \psi(\vec{x})\,d\vec{x} = 1$ and $\psi(\vec{x}) = 0$ for $|\vec{x}| \geq 1$, and let $\psi_\varepsilon(\vec{x}) = \varepsilon^{-n}\psi\left(\frac{1}{\varepsilon}\vec{x}\right)$. Then we define the ultraviolet cut-off field $\Phi_\varepsilon(\vec{x})$ by

$$\Phi_\varepsilon(\vec{x}) = \int \Phi(\vec{x} - \vec{y})\psi_\varepsilon(\vec{y})\,d\vec{y}. \tag{9.21}$$

Let now V be a real function of a real variable such that V is the Fourier transform of a bounded measure i.e. $V \in \mathcal{F}(\mathbb{R})$, we then define the space cut-off interaction V_Λ^ε, where Λ is a finite subset of $\mathbb{R}^n$, by

$$V_\Lambda^\varepsilon = \int_\Lambda V(\Phi_\varepsilon(\vec{x}))\,d\vec{x}. \tag{9.22}^2$$

Since

$$V(s) = \int e^{\mathrm{i}s\alpha}\,d\mu(\alpha), \tag{9.23}$$

(9.22) is defined to be the element associated with the Weyl-algebra given by

$$V_\Lambda^\varepsilon = \int_\Lambda \int e^{\mathrm{i}\alpha\Phi_\varepsilon(\vec{x})}\,d\mu(\alpha)\,d\vec{x} \tag{9.24}$$

or, by Definition 9.11,

$$V_\Lambda^\varepsilon = \int_\Lambda \int \epsilon\left(\alpha(2A)^{-\frac{1}{2}}\psi_\varepsilon^{\vec{x}}\right)\,d\mu(\alpha)\,d\vec{x}, \tag{9.25}$$

where $A = \sqrt{-\Delta + m^2}$ and $\psi_\varepsilon^{\vec{x}}(\vec{y}) = \psi_\varepsilon(\vec{y} - \vec{x})$. The integral (9.25) does not necessarily converge in the topology of the Weyl algebra or, for that-matter, in the natural C^*-topology of the Weyl algebra. However, in any representation induced by a state invariant under space translation, the representative of $\epsilon\left(\alpha(2A)^{-\frac{1}{2}}\psi_\varepsilon^{\vec{x}}\right)$ is strongly continuous in α and $\vec{x}$, since $\epsilon\left(\alpha(2A)^{-\frac{1}{2}}\psi_\varepsilon^{\vec{x}}\right) = e^{\mathrm{i}\alpha\Phi_\varepsilon(\vec{x})}$ is strongly continuous in α, because $\Phi_\varepsilon(\vec{x})$ is self adjoint and one has

$$e^{\mathrm{i}\alpha\Phi_\varepsilon(\vec{x})} = \underset{\vec{x}}{U}\, e^{\mathrm{i}\alpha\Phi_\varepsilon(\vec{0})}\,\underset{-\vec{x}}{U},$$

---

[2] As mentioned in the introduction, models with these interactions and their limit when the space cut-off is removed ($\Lambda \to \mathbb{R}^n$) have been studied before, see [30], [45] and references given therein.

where $U_{\vec{x}}$ is a strongly continuous representation of $\mathbb{R}^n$, the state being invariant under space translation. Hence, in any representation induced by a state which is space translation invariant, (9.25) exists as a strong Riemann integral and therefore $V_{\Lambda}^{\varepsilon}$ is represented there. Let now ρ be a state on the Weyl algebra which is space translation invariant. In the representation given by ρ we then have $V_{\Lambda}^{\varepsilon}$ represented, and we shall use the notation $V_{\Lambda}^{\varepsilon}$ for its representative. Assume now also that ρ is invariant under the free time isomorphism α_t^0. Then, in this representation α_t^0 is induced by a unitary group $\mathrm{e}^{-\mathrm{i}tH_0}$, where H_0 is the self adjoint infinitesimal generator for this unitary group. It follows easily from (9.25) that $V_{\Lambda}^{\varepsilon}$ is bounded, hence

$$H = H_0 + V_{\Lambda}^{\varepsilon} \tag{9.26}$$

is a self adjoint operator in the representation space. Let α_t be the automorphism on the bounded operators of the representation space induced by the unitary group $\mathrm{e}^{-\mathrm{i}tH}$. Let now ρ be any of the space–time invariant quasi free states of Theorem 9.1, we then have the following theorem.

Theorem 9.3. *Let ω_C be a quasi free state on the Weyl algebra for the free boson field on $\mathbb{R}^n$, invariant under space and time translations. Let $V \in \mathcal{F}(\mathbb{R})$ and define H_0 as the self adjoint operator generating α_t^0 in the representation given by ω_C. Let moreover H be the self adjoint operator in the representation space given by*

$$H = H_0 + \int\limits_{\Lambda} V(\Phi_\varepsilon(\vec{x})) \, \mathrm{d}\vec{x},$$

and let α_t be the automorphism induced by H on the algebra of bounded operators in the representation space. If $F_1, \ldots, F_n$ are in $\mathcal{F}(R)$, and $f_1, \ldots, f_n$ in $D\left(A^{-\frac{1}{2}}\right)$ and $t_1 \leq \ldots \leq t_n$, then

$$\mathrm{e}^{-\mathrm{i}\int\limits_{t_1}^{t_n}\int\limits_{\Lambda} V(\varphi_\varepsilon(\vec{x},t))\,\mathrm{d}\vec{x}\,\mathrm{d}t} \prod_{j=1}^{n} F_j\left(\int \varphi(\vec{x},t_j)f_j(\vec{x})\,\mathrm{d}\vec{x}\right)$$

is in $\mathcal{F}(D^*)$ and

$$\omega_C\left(\alpha_{t_1}(F_1(\Phi(f_1)))\alpha_{t_2}(F_2(\Phi(f_2)))\ldots\alpha_{t_n}(F_n(\Phi(f_n)))\right)$$

$$\overset{\Delta_C}{=} \int\limits_{\mathcal{H}} \mathrm{e}^{\mathrm{i}S(\varphi)}\mathrm{e}^{-\mathrm{i}\int_{t_1}^{t_n}\int_{\Lambda} V(\varphi_\varepsilon(\vec{x},t))\,\mathrm{d}\vec{x}\,\mathrm{d}t} \prod_{j=1}^{n} F_j\left(\int \varphi(\vec{x},t_j)f_j(\vec{x})\,\mathrm{d}\vec{x}\right) \mathrm{d}\varphi,$$

where

$$\varphi_\varepsilon(\vec{x},t) = \int \varphi(\vec{x}-\vec{y},t)\psi_\varepsilon(\vec{y})\,\mathrm{d}\vec{y}.$$

Proof. The proof of this theorem follows in the same way as the proof of Theorem 6.2 by series expansion and use of previous results of this section.

$\square$

Notes

This approach was first presented in the first edition of this book. It played a suggestive role in the formulation of the Euclidean fields with trigonometric interactions [88, 238]. It also directly inspired the treatment of the Chern–Simons model via path integrals [115, 116], see Sect. 10.5.5.

10

Some Recent Developments

10.1 The Infinite Dimensional Oscillatory Integral

In Chaps. 2 and 4 the "Fresnel integral", i.e. the oscillatory integral with quadratic phase function on a real separable (infinite dimensional) Hilbert space $\mathcal{H}$, is defined as a linear continuous functional on a Banach algebra of functions. We recall that a Fresnel integrable function is an element of $\mathcal{F}(\mathcal{H})$, the space of Fourier transforms of complex bounded variation measures on $\mathcal{H}$, and given a function $f \in \mathcal{F}(\mathcal{H})$, with $f(x) = \int e^{i(x,y)} d\mu_f(y)$, the corresponding Fresnel integral is defined by (2.9), i.e.

$$\widetilde{\int} e^{\frac{i}{2}|x|^2} f(x) dx = \int e^{-\frac{i}{2}|x|^2} d\mu(x).$$

In the 1980s Elworthy and Truman [223] proposed an alternative definition. The Fresnel integral is realized as an "infinite dimensional oscillatory integral", defined by means of a twofold limiting procedure. More precisely an oscillatory integral on an infinite dimensional Hilbert space is defined as the limit of a sequence of finite dimensional approximations, that are defined, according to an Hörmander proposal [278], as the limit of regularized, hence absolutely convergent Lebesgue integrals (see also, e.g., [220, 424]).

The study of oscillatory integrals on $\mathbb{R}^n$ of the form

$$\int_{\mathbb{R}^n} e^{i\Phi(x)} f(x) dx, \tag{10.1}$$

where $\Phi : \mathbb{R}^n \to \mathbb{R}$ and $f : \mathbb{R}^n \to \mathbb{C}$ are suitable smooth functions, is a classical topic, largely developed in connection with various problems in mathematics and physics. Well known examples of simple integrals of the above form are the Fresnel integrals of the theory of wave diffraction and Airy's integrals of the theory of rainbow. The theory of Fourier integral operators [278, 279, 364] also grew out of the investigation of oscillatory integrals. It allows the study of existence and regularity of a large class of elliptic and pseudoelliptic operators and provides constructive tools for the solutions of the corresponding (elliptic, parabolic and hyperbolic) equations.

If the function f is not absolutely integrable, then the integral (10.1) can be defined in the following way [278]:

Definition 10.1. *The oscillatory integral of a Borel function* $f : \mathbb{R}^n \to \mathbb{C}$ *with respect to a phase function* $\Phi : \mathbb{R}^n \to \mathbb{R}$ *is well defined if for each test function* $\varphi \in S(\mathbb{R}^n)$ *such that* $\varphi(0) = 1$ *the integral*

$$I_\epsilon(f, \varphi) := \int_{\mathbb{R}^n} e^{i\Phi(x)} f(x)\varphi(\epsilon x)\mathrm{d}x \tag{10.2}$$

exists for all $\epsilon > 0$ *and the limit* $\lim_{\epsilon \to 0} I_\epsilon(f, \varphi)$ *exists and is independent of* φ. *In this case the limit is called the oscillatory integral of* f *with respect to* Φ *and denoted by* $\int_{\mathbb{R}^n}^\circ e^{i\Phi(x)} f(x)\mathrm{d}x$.

In the particular case in which the phase function Φ is a non-degenerate quadratic form, in particular if Φ is proportional to the scalar product in $\mathbb{R}^n$, that is $\Phi(x) = \frac{(x,x)}{2\hbar} = \frac{|x|^2}{2\hbar}$ ($\hbar > 0$ being a strictly positive constant), it is convenient to include into the definition of the oscillatory integral the "normalization factor" $(2\pi i\hbar)^{n/2}$, which is fundamental in the extension of such a definition to the infinite dimensional case.

Definition 10.2. *A Borel function* $f : \mathbb{R}^n \to \mathbb{C}$ *is called (Fresnel-type) integrable if for each* $\varphi \in S(\mathbb{R}^n)$ *such that* $\varphi(0) = 1$ *the integral*

$$(2\pi i\hbar)^{-n/2} \int_{\mathbb{R}^n} e^{\frac{i}{2\hbar}|x|^2} f(x)\varphi(\epsilon x)\mathrm{d}x \tag{10.3}$$

exists for all $\epsilon > 0$ *and the limit*

$$\lim_{\epsilon \to 0} (2\pi i\hbar)^{-n/2} \int_{\mathbb{R}^n} e^{\frac{i}{2\hbar}|x|^2} f(x)\varphi(\epsilon x)\mathrm{d}x \tag{10.4}$$

exists and is independent of φ. *In this case the limit is called the Fresnel integral of* f *and denoted by*

$$\widetilde{\int_{\mathbb{R}^n}^\circ} e^{\frac{i}{2\hbar}|x|^2} f(x)\mathrm{d}x \tag{10.5}$$

Remark 10.1. One can easily verify that $\widetilde{\int_{\mathbb{R}^n}^\circ} e^{\frac{i}{2\hbar}|x|^2} f(x)\mathrm{d}x = 1$ if $f(x) = 1$ $\forall x \in \mathbb{R}^n$. In this sense the integral is normalized.

Remark 10.2. Definitions 10.1 and 10.2 are a generalization of the definition of normalized Fresnel integrals of Chap. 2, (2.9), as they allow, at least in principle, to define the oscillatory integral (10.1) for a more general class of phase functions Φ and integrands f than the quadratic forms and the Fourier transforms of measures. In fact this extension has been performed to include, in finite dimensions, all even degree polynomial phase functions [103] and, in infinite dimensions, quartic phase functions [105], see Sect. 10.2 below.

In [223] this definition is generalized to the case where $\mathbb{R}^n$ is replaced by a real separable infinite dimensional Hilbert space $(\mathcal{H}, (\ ,\ ))$:

Definition 10.3. *A Borel measurable function $f : \mathcal{H} \to \mathbb{C}$ is called $\mathcal{F}^\hbar$ integrable if for each sequence $\{P_n\}_{n \in \mathbb{N}}$ of projectors onto n-dimensional subspaces of $\mathcal{H}$, such that $P_n \leq P_{n+1}$ and $P_n \to I$ strongly as $n \to \infty$ (I being the identity operator in $\mathcal{H}$), the finite dimensional approximations of the oscillatory integral of f*

$$\mathcal{F}^\hbar_{P_n}(f) = \int_{P_n\mathcal{H}}^{\circ} e^{\frac{i}{2\hbar}|P_n x|^2} f(P_n x) \mathrm{d}(P_n x) \left(\int_{P_n\mathcal{H}}^{\circ} e^{\frac{i}{2\hbar}|P_n x|^2} \mathrm{d}(P_n x) \right)^{-1},$$

are well defined (in the sense of Definition 10.2) and the limit $\lim_{n\to\infty} \mathcal{F}^\hbar_{P_n}(f)$ exists and is independent of the sequence $\{P_n\}$.
In this case the limit is called the infinite dimensional oscillatory integral of f and is denoted by

$$\mathcal{F}^\hbar(f) = \widetilde{\int_{\mathcal{H}}^{\circ}} e^{\frac{i}{2\hbar}|x|^2} f(x) \mathrm{d}x.$$

The "concrete" description of the class of all $\mathcal{F}^\hbar$ integrable functions is still an open problem of harmonic analysis, even when $dim(\mathcal{H}) < \infty$. The following theorem shows that this class includes $\mathcal{F}(\mathcal{H})$, the class of Fresnel integrable functions in the sense of definitions (2.9) and (4.12) of Chaps. 2 and 4.

Theorem 10.1. *Let $L : \mathcal{H} \to \mathcal{H}$ be a self-adjoint trace class operator such that $(I - L)$ is invertible (I being the identity operator in $\mathcal{H}$). Let us assume that $f \in \mathcal{F}(\mathcal{H})$. Then the function $g : \mathcal{H} \to \mathbb{C}$ given by*

$$g(x) = e^{-\frac{i}{2\hbar}(x, Lx)} f(x), \qquad x \in \mathcal{H}$$

is $\mathcal{F}^\hbar$ integrable and the corresponding infinite dimensional oscillatory integral $\mathcal{F}^\hbar(g)$ is given by the following Cameron-Martin-Parseval type formula:

$$\widetilde{\int_{\mathcal{H}}^{\circ}} e^{\frac{i}{2\hbar}(x,(I-L)x)} f(x) \mathrm{d}x = (\det(I-L))^{-1/2} \int_{\mathcal{H}} e^{-\frac{i\hbar}{2}(x,(I-L)^{-1}x)} \mathrm{d}\mu_f(x) \quad (10.6)$$

where $\det(I - L) = |\det(I - L)|e^{-\pi i\, \mathrm{Ind}\,(I-L)}$ is the Fredholm determinant of the operator $(I - L)$, $|\det(I - L)|$ its absolute value and $\mathrm{Ind}((I - L))$ is the number of negative eigenvalues of the operator $(I - L)$, counted with their multiplicity.

Proof. Given a sequence $\{P_n\}_{n \in \mathbb{N}}$ of projectors onto n-dimensional subspaces of $\mathcal{H}$, such that $P_n \leq P_{n+1}$ and $P_n \to I$ strongly as $n \to \infty$ (I being the identity operator in $\mathcal{H}$), the finite dimensional approximations of the oscillatory integral $\widetilde{\int_{\mathcal{H}}^{\circ}} e^{\frac{i}{2\hbar}(x,(I-L)x)} f(x) \mathrm{d}x$ are equal to:

$$\int_{P_n\mathcal{H}}^{\circ} e^{\frac{i}{2\hbar}|P_n x|^2} e^{-\frac{i}{2\hbar}(P_n x, L P_n x)} f(P_n x) \mathrm{d}(P_n x) \left(\int_{P_n\mathcal{H}}^{\circ} e^{\frac{i}{2\hbar}|P_n x|^2} \mathrm{d}(P_n x) \right)^{-1}.$$

$$(10.7)$$

Let $L_n : P_n\mathcal{H} \to P_n\mathcal{H}$ be the operator on $P_n\mathcal{H}$ given by $L_n := P_n L P_n$. As $(I - L)$ is invertible, it is easy to see that for n sufficiently large the operator $(I - L_n)$ on $P_n\mathcal{H}$ is invertible and by Parseval's formula the expression (10.7) is equal to

$$(\det(I - L_n))^{-1/2} \int_{P_n\mathcal{H}} \mathrm{e}^{-\frac{\mathrm{i}\hbar}{2}(P_n x,(I-L_n)^{-1}P_n x)} \mathrm{d}\mu_f(P_n x). \tag{10.8}$$

By letting $n \to \infty$, $L_n \to L$ in trace norm and expression (10.8) converges to the right hand side of (10.6). For more details see [69, 223]. □

Remark 10.3. We have $\int_{\mathcal{H}}^{\circ} \mathrm{e}^{\frac{\mathrm{i}}{2\hbar}(x,(I-L)x)} f(x)\mathrm{d}x = 1$ for $f(x) = 1 \ \forall x \in \mathcal{H}$ only when $L = 0$. In this sense the integral is not normalized.

On the other hand the normalization factor $\left(\int_{P_n\mathcal{H}}^{\circ} \mathrm{e}^{\frac{\mathrm{i}}{2\hbar}|P_n x|^2} \mathrm{d}(P_n x) \right)^{-1}$ in the finite dimensional approximations of the infinite dimensional oscillatory integral is fundamental. In fact it makes the definition of the infinite dimensional oscillatory integral coherent with the definition of the oscillatory integral on $\mathbb{R}^n$. Indeed it is possible to generalize Proposition 2.2, in other words it is possible to prove that if $f \in \mathcal{F}(\mathcal{H})$ is a finitely based function, i.e. there exists a finite dimensional orthogonal projection P in $\mathcal{H}$ such that $f(x) = f(Px)$ for all $x \in \mathcal{H}$, then

$$\int_{\mathcal{H}}^{\circ} \mathrm{e}^{\frac{\mathrm{i}}{2\hbar}|x|^2} f(x)\mathrm{d}x = \int_{P\mathcal{H}}^{\circ} \mathrm{e}^{\frac{\mathrm{i}}{2\hbar}|x|^2} f(x)\mathrm{d}x,$$

where the left hand side denotes an infinite dimensional oscillatory integral (Definition 10.3) and the right hand side the oscillatory integral on the finite dimensional space $P\mathcal{H}$ (Definition 10.2).

Results similar to those obtained in Chap. 4 (the Fresnel integral relative to a non singular quadratic form) can be obtained by introducing in the finite dimensional approximations a suitable normalization constant. Indeed given a self-adjoint invertible operator B on $\mathcal{H}$, it is possible to define the normalized infinite dimensional oscillatory integral with respect to B.

Definition 10.4. A Borel function $f : \mathcal{H} \to \mathbb{C}$ is called $\mathcal{F}_B^\hbar$ integrable if for each sequence $\{P_n\}_{n\in\mathbb{N}}$ of projectors onto n-dimensional subspaces of $\mathcal{H}$, such that $P_n \leq P_{n+1}$ and $P_n \to I$ strongly as $n \to \infty$ (I being the identity operator in $\mathcal{H}$) the finite dimensional approximations

$$\int_{P_n\mathcal{H}}^{\circ} \mathrm{e}^{\frac{\mathrm{i}}{2\hbar}(P_n x, B P_n x)} f(P_n x)\mathrm{d}(P_n x),$$

are well defined and the limit

$$\lim_{n\to\infty} (\det P_n B P_n)^{\frac{1}{2}} \int_{P_n\mathcal{H}}^{\circ} \mathrm{e}^{\frac{\mathrm{i}}{2\hbar}(P_n x, B P_n x)} f(P_n x)\mathrm{d}(P_n x) \tag{10.9}$$

exists and is independent of the sequence $\{P_n\}$.

In this case the limit is called the normalized oscillatory integral of f with respect to B and is denoted by

$$\widetilde{\int_{\mathcal{H}}^{B}} e^{\frac{i}{2\hbar}(x,Bx)} f(x) \mathrm{d}x$$

Again, given a function $f \in \mathcal{F}(\mathcal{H})$, it is possible to prove that f is $\mathcal{F}_B^{\hbar}$ integrable and the corresponding normalized infinite dimensional oscillatory integral can be computed by means of a formula similar to (10.6):

Theorem 10.2. *Let us assume that $f \in \mathcal{F}(\mathcal{H})$. Then f is $\mathcal{F}_B^{\hbar}$ integrable and the corresponding normalized oscillatory integral is given by the following Cameron–Martin–Parseval type formula:*

$$\widetilde{\int_{\mathcal{H}}^{B}} e^{\frac{i}{2\hbar}(x,Bx)} f(x) \mathrm{d}x = \int_{\mathcal{H}} e^{-\frac{i\hbar}{2}(x,B^{-1}x)} \mathrm{d}\mu_f(x). \tag{10.10}$$

Note that if we substitute into the latter the function $f = 1$, we have $\widetilde{\int^{B}} e^{\frac{i}{2\hbar}(x,Bx)} f(x) dx = 1$. For this reason the integral is called "normalized".
The latter theorem shows that in the infinite dimensional case the normalization constant in the finite dimensional approximations plays a crucial role and Definitions 10.3 and 10.4 are not equivalent. Indeed Theorem 10.2 makes sense even if the operator $L := I - B$ is not trace class (in which case the Fredholm determinant $\det(I - B)$ cannot be defined).
In fact it is possible to introduce different normalization constants in the finite dimensional approximations and the properties of the corresponding infinite dimensional oscillatory integrals are related to the trace properties of the operator associated to the quadratic part of the phase function [70]. More precisely, for any $p \in \mathbb{N}$, let us consider the Schatten class $\mathcal{T}_p(\mathcal{H})$ of bounded linear operators L in $\mathcal{H}$ such that

$$\|L\|_p = (\mathrm{Tr}(L^*L)^{p/2})^{1/p}$$

is finite. $(\mathcal{T}_p(\mathcal{H}), \|\cdot\|_p)$ is a Banach space. For any $p \in \mathbb{N}$, $p \geq 2$ and $L \in \mathcal{T}_p(\mathcal{H})$ one defines the regularized Fredholm determinant $\det_{(p)} : I + \mathcal{T}_p(\mathcal{H}) \to \mathbb{R}$:

$$\det_{(p)}(I + L) = \det\left((I + L)\exp\sum_{j=1}^{p-1}\frac{(-1)^j}{j}L^j\right), \qquad L \in \mathcal{T}_p(\mathcal{H}),$$

where det denotes the usual Fredholm determinant, which is well defined as it is possible to prove that the operator $(I + L)\exp\sum_{j=1}^{p-1}\frac{(-1)^j}{j}L^j - I$ is trace class [412]. In particular $\det_{(2)}$ is called Carleman determinant.
For $p \in \mathbb{N}$, $p \geq 2$, $L \in \mathcal{T}_1(\mathcal{H})$, let us define the normalized quadratic form on $\mathcal{H}$:

$$N_p(L)(x) = (x, Lx) - i\hbar\mathrm{Tr}\sum_{j=1}^{p-1}\frac{L^j}{j}, \qquad x \in \mathcal{H}. \tag{10.11}$$

Again, for $p \in \mathbb{N}$, $p \geq 2$, let us define the *class p normalized oscillatory integral*.

Definition 10.5. *Let $p \in \mathbb{N}$, $p \geq 2$, L a bounded linear operator in $\mathcal{H}$, $f : \mathcal{H} \to \mathbb{C}$ a Borel measurable function. The class p normalized oscillatory integral of the function f with respect to the operator L is well defined if for each sequence $\{P_n\}_{n \in \mathbb{N}}$ of projectors onto n-dimensional subspaces of $\mathcal{H}$, such that $P_n \leq P_{n+1}$ and $P_n \to I$ strongly as $n \to \infty$ (I being the identity operator in $\mathcal{H}$) the finite dimensional approximations*

$$\widetilde{\int_{P_n \mathcal{H}}^{\circ}} e^{\frac{i}{2\hbar}|x|^2} e^{-\frac{i}{2\hbar} N_p(P_n L P_n)(P_n x)} f(P_n x) d(P_n x), \qquad (10.12)$$

are well defined and the limit

$$\lim_{n \to \infty} \widetilde{\int_{P_n \mathcal{H}}^{\circ}} e^{\frac{i}{2\hbar}|x|^2} e^{-\frac{i}{2\hbar} N_p(P_n L P_n)(P_n x)} f(P_n x) d(P_n x) \qquad (10.13)$$

exists and is independent of the sequence $\{P_n\}$.
In this case the limit is denoted by

$$\mathcal{I}_{p,L}(f) = \widetilde{\int_{\mathcal{H}}^{p}} e^{\frac{i}{2\hbar}|x|^2} e^{-\frac{i}{2\hbar}(x, Lx)} f(x) dx.$$

If L is not a trace class operator, then the quadratic form (10.11) is not well defined. Nevertheless expression (10.12) still makes sense thanks to the fact that all the functions under the integral are restricted to finite dimensional subspaces. Moreover the limit (10.13) can make sense, as the following result shows [70].

Theorem 10.3. *Let us assume that $f \in \mathcal{F}(\mathcal{H})$, $L = L^*$, $L \in \mathcal{T}_p(\mathcal{H})$ and $\det_{(p)}(I - L) \neq 0$. Then the class-p normalized oscillatory integral of the function f with respect to the operator L exists and is given by the following Cameron–Martin–Parseval type formula:*

$$\widetilde{\int_{\mathcal{H}}^{p}} e^{\frac{i}{2\hbar}|x|^2} e^{-\frac{i}{2\hbar}(x, Lx)} f(x) dx = [\det_{(p)}(I - L)]^{-1/2} \int_{\mathcal{H}} e^{-\frac{i\hbar}{2}(x, (I-L)^{-1}x)} d\mu_f(x).$$
$$(10.14)$$

Similarly to what is done in Chaps. 3 and 5, it is possible to prove that, under suitable assumptions on the initial datum φ, the solution to the Schrödinger equation for an anharmonic oscillator potential

$$\begin{cases} i\hbar \frac{\partial}{\partial t} \psi = -\frac{\hbar^2}{2m} \triangle \psi + (\frac{1}{2} x A^2 x + V(x)) \psi \\ \psi(0, x) = \varphi(x), \quad x \in \mathbb{R}^d \end{cases} \qquad (10.15)$$

(where $A^2 \geq 0$ is a positive $d \times d$ matrix with constant elements) can be represented by a well defined infinite dimensional oscillatory integral on the Hilbert space $(\mathcal{H}_t, (\,,\,))$ of real continuous functions $\gamma(\tau)$ from $[0, t]$ to $\mathbb{R}^d$ such that

$\frac{d\gamma}{d\tau} \in L_2([0, t]; \mathbb{R}^d)$ and $\gamma(t) = 0$ with inner product $(\gamma_1, \gamma_2) = \int\limits_0^t \frac{d\gamma_1}{d\tau} \cdot \frac{d\gamma_2}{d\tau} d\tau$.

From now on we will assume for notational simplicity that $m = 1$, but the whole discussion can be generalized to arbitrary $m > 0$.

By considering the unique self-adjoint operator L on $\mathcal{H}_t$, uniquely determined by the quadratic form

$$(\gamma_1, L\gamma_2) \equiv \int_0^t \gamma_1(\tau)A^2\gamma_2(\tau)d\tau,$$

and the function $v : \mathcal{H}_t \to \mathbb{C}$

$$v(\gamma) \equiv \int_0^t V(\gamma(\tau) + x)d\tau + 2xA^2 \int_0^t \gamma(\tau)d\tau, \qquad \gamma \in \mathcal{H}_t, x \in \mathbb{R}^d,$$

Feynman's heuristic formula

$$\text{`` } const \int_{\{\gamma | \gamma(t) = x\}} e^{\frac{i}{\hbar} \int_0^t (\frac{1}{2}\dot{\gamma}(\tau)^2 - \frac{1}{2}\gamma(\tau)A^2\gamma(\tau) - V(\gamma(\tau)))d\tau} \varphi(\gamma(0))d\gamma \text{ ''}$$

can be interpreted as the infinite dimensional oscillatory integral on $\mathcal{H}_t$ (in the sense of Definition 10.3).

$$\widetilde{\int_{\mathcal{H}_t}^{\circ}} e^{\frac{i}{2\hbar}(\gamma, (I-L)\gamma)} e^{-\frac{i}{\hbar}v(\gamma)} \varphi(\gamma(0) + x)d\gamma. \tag{10.16}$$

One can easily verify that the operator $L : \mathcal{H}_t \to \mathcal{H}_t$ is given on the vectors $\gamma \in \mathcal{H}_t$ by

$$L\gamma(\tau) = \int_\tau^t \int_0^r A^2\gamma(s)dsdr.$$

By analyzing the spectrum of L (see [223] for more details) it is easy to see that L is trace class and $I - L$ is invertible. The following holds:

Theorem 10.4. *Let $\varphi \in \mathcal{F}(\mathbb{R}^d) \cap L^2(\mathbb{R}^d)$ and let $V \in \mathcal{F}(\mathbb{R}^d)$. Then the function $f_x : \mathcal{H}_t \to \mathbb{C}$, $x \in \mathbb{R}^d$, given by*

$$f_x(\gamma) = e^{-\frac{i}{\hbar}v(\gamma)}\varphi(\gamma(0) + x)$$

is the Fourier transform of a complex bounded variation measure μ_{f_x} on $\mathcal{H}_t$ and the infinite dimensional Fresnel integral of the function $g_x(\gamma) = e^{-\frac{i}{2\hbar}(\gamma, L\gamma)}f_x(\gamma)$

$$\widetilde{\int_{\mathcal{H}_t}^{\circ}} e^{\frac{i}{2\hbar}(\gamma, (I-L)\gamma)} e^{-\frac{i}{\hbar}v(\gamma)} \varphi(\gamma(0) + x)d\gamma.$$

is well defined (in the sense of Definition 10.3) and it is equal to

$$\det(I - L)^{-1/2} \int_{\mathcal{H}_t} e^{-\frac{i\hbar}{2}(\gamma, (I-L)^{-1}\gamma)} d\mu_{f_x}(\gamma).$$

Moreover it is a representation of the solution of equation (10.15) evaluated at $x \in \mathbb{R}^d$ at time t.

For a proof see [223].

There is an interesting difference between (4.12), the definition of the Fresnel integral with respect to a non singular quadratic form, and equation (10.6), the Parseval type equality for the infinite dimensional oscillatory integral. In the latter one finds the multiplicative factor $\det(I - L)^{-1/2}$ (the Fredholm determinant) in front of the absolutely convergent integral on the right hand side. As a consequence, equation (5.37) for the Fresnel integral representation of the solution of the Schrödinger equation (10.15),

$$\psi(x,t) = |\cos At|^{-\frac{1}{2}} \int\limits_{\gamma(t)=x}^{\sim} e^{\frac{1}{2}\int_0^t(\dot\gamma^2 - \gamma A^2\gamma)d\tau} \, e^{-i\int_0^t V(\gamma(\tau))d\tau} \varphi(\gamma(0))d\gamma$$

(where the r.h.s. denotes a Fresnel integral in the sense of Definition 4.12) can accordingly be replaced by

$$\psi(x,t) = \int\limits_{\gamma(t)=x}^{\widetilde{\quad\circ\quad}} e^{\frac{1}{2}\int_0^t(\dot\gamma^2 - \gamma A^2\gamma)d\tau} \, e^{-i\int_0^t V(\gamma(\tau))d\tau} \varphi(\gamma(0))d\gamma$$

(where the r.h.s. denotes an infinite dimensional oscillatory integral in the sense of Definition 10.3). As one can see the latter formula does not contain the "strange looking" multiplicative factor $|\cos At|^{-\frac{1}{2}}$.

It is interesting to note that the sequential approach of Definition 10.3 is not so different from Feynman's original derivation of his famous heuristic formula (1.13). In fact let us consider the sequence of partitions π_n of the interval $[0, t]$ into n subintervals of amplitude $\epsilon \equiv t/n$:

$$t_0 = 0, t_1 = \epsilon, \ldots, t_i = i\epsilon, \ldots, t_n = n\epsilon = t.$$

To each π_n let us associate a projector $P_n : \mathcal{H}_t \to \mathcal{H}_t$ onto a finite dimensional subspace of $\mathcal{H}_t$, consisting of piecewise polygonal paths. In other words each projector P_n acts on a path $\gamma \in \mathcal{H}_t$ in the following way:

$$P_n(\gamma)(\tau) = \sum_{i=1}^n \chi_{[t_{i-1},t_i]}(\tau)\left(\gamma(t_{i-1}) + \frac{(\gamma(t_i) - \gamma(t_{i-1}))}{t_i - t_{i-1}}(\tau - t_{i-1})\right).$$

By considering the infinite dimensional oscillatory integral on $\mathcal{H}_t$ of the function $f_x(\gamma) = e^{-\frac{i}{\hbar}\int_0^t V(\gamma(\tau)+x)d\tau}\varphi(\gamma(0) + x)$, one has

$$\int_{\mathcal{H}_t}^{\widetilde{\quad\circ\quad}} e^{\frac{i}{2\hbar}(\gamma,\gamma)} f_x(\gamma)d\gamma$$

$$= \lim_{n\to\infty} \int_{P_n\mathcal{H}_t}^{\widetilde{\quad\circ\quad}} e^{\frac{i}{2\hbar}(P_n\gamma, P_n\gamma)} f_x(P_n\gamma)dP_n\gamma \left[\int_{P_n\mathcal{H}_t}^{\widetilde{\quad\circ\quad}} e^{\frac{i}{2\hbar}(P_n\gamma, P_n\gamma)}dP_n\gamma\right]^{-1}$$

$$= \lim_{n\to\infty} (2\pi i\hbar t/n)^{-\frac{nd}{2}} \int_{\mathbb{R}^{nk}} e^{-\frac{i}{\hbar}S_t(x_k,\ldots,x_0)}\varphi(\gamma(0) + x)\prod_{i=0}^{n-1} d\gamma(t_i) \quad (10.17)$$

and one can recognize a formula analogous to (1.10) proposed by Feynman in his original work. See [449] for the details.

Remark 10.4. Another reference to the sequential approach is [236] (which contain in particular a discussion of continuous quantum observations). See also Sect. 10.4.3 for a sequential approach through "classical paths" instead of "polygonal paths".

10.2 Feynman Path Integrals for Polynomially Growing Potentials

As we have seen in the previous section, the Fresnel integral approach allows to give a rigorous mathematical meaning to the Feynman path integral representation of the solution of the Schrödinger equation if the potential V is of the type "quadratic plus bounded perturbation". Indeed, in order to define the infinite dimensional Fresnel integral, the perturbation to the harmonic oscillator potential has to belong to $\mathcal{F}(\mathbb{R}^d)$, so that in particular it is bounded. An extension to unbounded potentials which are Laplace transforms of bounded measures has been developed in [73, 336] by means of the analytic continuation approach resp. by means of white noise analysis. It includes some exponentially growing potentials but does not cover the case of potentials which are polynomials of degree larger than 2. In fact the problem for such polynomial potentials is not simple, as it has been proved [466] that in one dimension, if the potential is time independent and super-quadratic in the sense that $V(x) \geq C(1 + |x|)^{2+\epsilon}$ at infinity, $C > 0$ and $\epsilon > 0$, then, as a function of (t, x, y), the fundamental solution $E(t, 0, x, y)$ of the time dependent Schrödinger equation is nowhere C^1.

As we have seen in the previous section, the Definition 10.3 of the infinite dimensional oscillatory integral is more flexible than definitions (2.9) and (4.12). In fact it allows, at least in principle, to enlarge the class of "integrable functions". This fact is used in [104, 105] for providing a direct rigorous Feynman path integral definition for the solution of the Schrödinger equation for an anharmonic oscillator potential $V(x) = \frac{1}{2}xA^2x + \lambda x^4$, $\lambda > 0$, (written for $d = 1$) without using the "indirect" tool of analytic continuation from a representation of the heat equation as for instance in [383, 217, 142, 441, 440, 207]. The first step is the definition and the computation of the oscillatory integral

$$\int_{\mathbb{R}^n}^{o} e^{\frac{i}{\hbar}\Phi(x)} f(x)dx,$$

for the case where the phase function $\Phi(x) = P(x)$ is an arbitrary even polynomial with positive leading coefficient. In this case a generalization of Theorem 10.1 can be proved. In complete analogy with Fresnel integrals, one uses the duality introduced by the Parseval type equality. The problem is the computation and the study of the Fourier transform of the distribution $e^{\frac{i}{\hbar}P(x)}$. The main tool is the following lemma, which can be proved by using the analyticity of the function $z \mapsto e^{kz+\frac{i}{\hbar}P(z)}$, $z \in \mathbb{C}^n$, $k \in \mathbb{R}^n$ and a change of integration contour (see [103]).

Lemma 10.1. *Let* $P : \mathbb{R}^n \to \mathbb{R}$ *be a* $2M$*-degree polynomial with positive leading coefficient. Then the Fourier transform of the distribution* $e^{\frac{i}{\hbar}P(x)}$:

$$\tilde{F}(k) = \int_{\mathbb{R}^n} e^{ik\cdot x} e^{\frac{i}{\hbar}P(x)} dx, \qquad \hbar \in \mathbb{R} \setminus \{0\} \qquad (10.18)$$

is an entire bounded function and admits the following representations:

$$\tilde{F}(k) = e^{in\pi/(4M)} \int_{\mathbb{R}^n} e^{ie^{i\pi/(4M)}k\cdot x} e^{\frac{i}{\hbar}P(e^{i\pi/(4M)}x)} dx, \qquad \text{for } \hbar > 0 \quad (10.19)$$

or

$$\tilde{F}(k) = e^{-in\pi/(4M)} \int_{\mathbb{R}^n} e^{ie^{-i\pi/(4M)}k\cdot x} e^{\frac{i}{\hbar}P(e^{-i\pi/(4M)}x)} dx, \qquad \text{for } \hbar < 0 \quad (10.20)$$

Remark 10.5. The integrals on the r.h.s. of (10.19) and (10.20) are absolutely convergent thanks to the fast decreasing behavior of $e^{\frac{i}{\hbar}P(e^{i\pi/(4M)}x)}$ resp. $e^{\frac{i}{\hbar}P(e^{-i\pi/(4M)}x)}$ when $|x| \to \infty$ (for $\hbar > 0$ resp. $\hbar < 0$).

Lemma 10.1 allows the following generalization of Theorem 10.1 to the case where the phase function is equal to an even degree polynomial P on $\mathbb{R}^n$:

Theorem 10.5. *Let* $P : \mathbb{R}^n \to \mathbb{R}$ *be a* $2M$*-degree polynomial with positive leading coefficient, and let* $f \in \mathcal{F}(\mathbb{R}^n)$, $f = \hat{\mu}_f$. *Then the oscillatory integral*

$$\widetilde{\int_{\mathbb{R}^n}} e^{\frac{i}{\hbar}P(x)} f(x) dx, \qquad \hbar \in \mathbb{R} \setminus \{0\}$$

is well defined and it is given by the formula of Parseval's type:

$$\widetilde{\int_{\mathbb{R}^n}} e^{\frac{i}{\hbar}P(x)} f(x) dx = \int \tilde{F}(k) d\mu_f(k), \qquad (10.21)$$

where $\tilde{F}(k)$ *defined by* (10.18) *and is given by* (10.19) *resp.* (10.20)
     The integral on the r.h.s. of (10.21) is absolutely convergent (hence it can be understood in Lebesgue sense).

It is particularly interesting to examine the case in which

$$P(x) = \frac{1}{2}x(I - L)x - \lambda B(x, x, x, x),$$

where $\hbar > 0$, $\lambda \leq 0$, I, L are $n \times n$ matrices, I being the identity, $(I - L)$ is symmetric and strictly positive and $B : \mathbb{R}^n \times \mathbb{R}^n \times \mathbb{R}^n \times \mathbb{R}^n \to \mathbb{R}$ is a completely symmetric and positive fourth order covariant tensor on $\mathbb{R}^n$.

Lemma 10.2. *Under the assumptions above the Fourier transform $\tilde{F}(k)$ of the distribution $\frac{e^{\frac{i}{\hbar}x\cdot(I-L)x}}{(2\pi i\hbar)^{n/2}}e^{\frac{-i\lambda}{\hbar}B(x,x,x,x)}$, i.e.:*

$$\tilde{F}(k) = \int_{\mathbb{R}^n} e^{ik\cdot x}\frac{e^{\frac{i}{2\hbar}x\cdot(I-L)x}}{(2\pi i\hbar)^{n/2}}e^{\frac{-i\lambda}{\hbar}B(x,x,x,x)}\mathrm{d}x, \qquad (10.22)$$

is a bounded complex-valued entire function on $\mathbb{R}^n$ admitting the following representation

$$\tilde{F}(k) = \int_{\mathbb{R}^n} e^{ie^{i\pi/4}k\cdot x}\frac{e^{-\frac{1}{2\hbar}x\cdot(I-L)x}}{(2\pi\hbar)^{n/2}}e^{\frac{i\lambda}{\hbar}B(x,x,x,x)}\mathrm{d}x =$$

$$= \mathbb{E}[e^{ie^{i\pi/4}k\cdot x}e^{\frac{i\lambda}{\hbar}A(x,x,x,x)}e^{\frac{1}{2\hbar}x\cdot Bx}], \quad (10.23)$$

where $\mathbb{E}$ denotes the expectation value with respect to the centered Gaussian measure μ_G on $\mathbb{R}^n$ with covariance operator $\hbar I$ (i.e. $\mu_G(\mathrm{d}x) = \frac{e^{-\frac{1}{2\hbar}|x|^2}}{(2\pi\hbar)^{n/2}}\mathrm{d}x$).

Theorem 10.6. *("Parseval equality") Let $f \in \mathcal{F}(\mathbb{R}^n)$, $f = \hat{\mu}_f$. Then, under the assumptions above, the generalized Fresnel integral*

$$\widetilde{\int_{\mathbb{R}^n}}^{\circ} e^{\frac{i}{2\hbar}x\cdot(I-L)x}e^{\frac{-i\lambda}{\hbar}B(x,x,x,x)}f(x)\mathrm{d}x$$

is well defined and it is given by:

$$\widetilde{\int_{\mathbb{R}^n}}^{\circ} e^{\frac{i}{2\hbar}x\cdot(I-L)x}e^{\frac{-i\lambda}{\hbar}B(x,x,x,x)}f(x)\mathrm{d}x = \int_{\mathbb{R}^n}\tilde{F}(k)\mathrm{d}\mu_f(k), \qquad (10.24)$$

where $\tilde{F}(k)$ is given by (10.23).

Moreover if μ_f is such that the integral $\int e^{-\frac{\sqrt{2}}{2}kx}\mathrm{d}|\mu_f|(k)$ is convergent for all $x \in \mathbb{R}^n$ and the positive function $g : \mathbb{R}^n \to \mathbb{R}^n$, defined by

$$g(x) = e^{\frac{1}{2\hbar}x\cdot Lx}\int e^{-\frac{\sqrt{2}}{2}kx}\mathrm{d}|\mu_f|(k),$$

is summable with respect to the centered Gaussian measure on $\mathbb{R}^n$ with covariance $\hbar I$, then f extends to an analytic function on $\mathbb{C}^n$ and the corresponding generalized Fresnel integral is given by:

$$\widetilde{\int_{\mathbb{R}^n}}^{\circ} e^{\frac{i}{2\hbar}x\cdot(I-L)x}e^{\frac{-i\lambda}{\hbar}B(x,x,x,x)}f(x)\mathrm{d}x = \mathbb{E}[e^{\frac{i\lambda}{\hbar}B(x,x,x,x)}e^{\frac{1}{2\hbar}x\cdot Lx}f(e^{i\pi/4}x)].$$

$$(10.25)$$

The technique used in the proof of Lemma 10.2 is similar to that in Lemma 10.1: again one uses the analyticity of the integrand and a change of integration contour. However in (10.23) the convergence of the integral is not given by the leading term of the polynomial phase function, as in (10.19) and (10.20), but is given by the Gaussian density $e^{-\frac{1}{2\hbar}x\cdot(I-L)x}$. This allows the generalization of the result to the infinite dimensional case. Indeed given a real separable infinite dimensional Hilbert space $\mathcal{H}$ with inner product $(\,,\,)$ and norm $|\,|$, let ν be the finitely additive cylinder measure on $\mathcal{H}$, defined by its characteristic functional $\hat{\nu}(x) = e^{-\frac{\hbar}{2}|x|^2}$. Let $\|\,\|$ be a "measurable norm" on $\mathcal{H}$ (in the sense of Gross [257]), that is $\|\,\|$ is such that for every $\epsilon > 0$ there exist a finite-dimensional projection $P_\epsilon : \mathcal{H} \to \mathcal{H}$, such that for all $P \perp P_\epsilon$ one has $\nu(\{x \in \mathcal{H} |\, \|P(x)\| > \epsilon\}) < \epsilon$, where P and P_ϵ are called orthogonal ($P \perp P_\epsilon$) if their ranges are orthogonal in $(\mathcal{H}, (\,,\,))$. One can easily verify that $\|\,\|$ is weaker than $|\,|$. Denoting by $\mathcal{B}$ the completion of $\mathcal{H}$ in the $\|\,\|$-norm and by i the continuous inclusion of $\mathcal{H}$ in $\mathcal{B}$, one can prove that $\mu \equiv \nu \circ i^{-1}$ is a countably additive Gaussian measure on the Borel subsets of $\mathcal{B}$. The triple $(i, \mathcal{H}, \mathcal{B})$ is called an *abstract Wiener space* [257, 337]. Given $y \in \mathcal{B}^*$ one can easily verify that the restriction of y to $\mathcal{H}$ is continuous on $\mathcal{H}$, so that one can identify $\mathcal{B}^*$ as a subset of $\mathcal{H}$ and each element $y \in \mathcal{B}^*$ can be regarded as a random variable $n(y)$ on $(\mathcal{B}, \mu)$. Given an orthogonal projection P in $\mathcal{H}$, with

$$P(x) = \sum_{i=1}^{n}(e_i, x)e_i$$

for some orthonormal $e_1, \ldots, e_n \in \mathcal{H}$, the stochastic extension $\tilde{P}$ of P on $\mathcal{B}$ is well defined by

$$\tilde{P}(\,\cdot\,) = \sum_{i=1}^{n} n(e_i)(\,\cdot\,)e_i.$$

Given a function $f : \mathcal{H} \to \mathcal{B}_1$, where $(\mathcal{B}_1, \|\,\|_{\mathcal{B}_1})$ is another real separable Banach space, the stochastic extension $\tilde{f}$ of f to $\mathcal{B}$ exists if the functions $f \circ \tilde{P} : \mathcal{B} \to \mathcal{B}_1$ converge to $\tilde{f}$ in probability with respect to μ as P converges strongly to the identity in $\mathcal{H}$.

Let $B : \mathcal{H} \times \mathcal{H} \times \mathcal{H} \times \mathcal{H} \to \mathbb{R}$ be a completely symmetric positive covariant tensor operator on $\mathcal{H}$ such that the map $V : \mathcal{H} \to \mathbb{R}^+$, $x \mapsto V(x) \equiv B(x, x, x, x)$ is continuous in the $\|\,\|$ norm. As a consequence V is continuous in the $|\,|$-norm, moreover it can be extended by continuity to a random variable $\bar{V}$ on $\mathcal{B}$, with $\bar{V}|_{\mathcal{H}} = V$. Moreover given a self-adjoint trace class operator $L : \mathcal{H} \to \mathcal{H}$, the quadratic form on $\mathcal{H} \times \mathcal{H}$:

$$x \in \mathcal{H} \mapsto (x, Lx)$$

can be extended to a random variable on $\mathcal{B}$, denoted again by $(\,\cdot\,, L\,\cdot\,)$. In this setting one can prove the following generalization of Theorem 10.6 [105].

Theorem 10.7. *Let L be self-adjoint trace class, $(I - L)$ strictly positive, $\lambda \le 0$ and $f \in \mathcal{F}(\mathcal{H})$, $f \equiv \hat{\mu}_f$, and let us assume that the bounded variation measure μ_f satisfies the following assumption*

$$\int_{\mathcal{H}} e^{\frac{\hbar}{4}(x,(I-L)^{-1}x)} \mathrm{d}|\mu_f|(x) < +\infty. \tag{10.26}$$

Then the infinite dimensional oscillatory integral

$$\widetilde{\int_{\mathcal{H}}^{\circ}} e^{\frac{i}{2\hbar}(x,(I-L)x)} e^{-i\frac{\lambda}{\hbar}B(x,x,x,x)} f(x) \mathrm{d}x \tag{10.27}$$

exists and is given by:

$$\int_{\mathcal{H}} \mathbb{E}[e^{in(k)(\omega)e^{i\pi/4}} e^{\frac{1}{2\hbar}(\omega,L\omega)} e^{i\frac{\lambda}{\hbar}\bar{V}(\omega)}] \mathrm{d}\mu_f(k)$$

It is also equal to :

$$\mathbb{E}[e^{i\frac{\lambda}{\hbar}\bar{V}(\omega)} e^{\frac{1}{2\hbar}(\omega,L\omega)} f(e^{i\pi/4}\omega)], \tag{10.28}$$

where $\mathbb{E}$ denotes the expectation value with respect to the Gaussian measure μ on $\mathcal{B}$.

Such a result allows for an extension of the class of potentials for which an infinite dimensional oscillatory integral representation of the solution of the corresponding Schrödinger equation can be defined. Let us consider the Schrödinger equation

$$i\hbar \frac{\partial}{\partial t} \psi = H\psi \tag{10.29}$$

on $L^2(\mathbb{R}^d)$ for an anharmonic oscillator Hamiltonian H of the following form:

$$H = -\frac{\hbar^2}{2}\triangle + \frac{1}{2}xA^2x + \lambda C(x,x,x,x), \tag{10.30}$$

where C is a completely symmetric positive fourth order covariant tensor on $\mathbb{R}^d$, A is a positive symmetric $d \times d$ matrix, $\lambda \geq 0$ a positive constant. It is well known, see [395], that H is essentially self-adjoint on $C_0^\infty(\mathbb{R}^d)$. Theorem 10.7 allows to give a well defined mathematical meaning to the "Feynman path integral" representation of the solution of (10.29) with initial datum $\varphi \in L_2(\mathbb{R}^d)$:

$$\psi(t,x) =$$
$$\text{``} \int_{\gamma(t)=x} e^{\frac{i}{\hbar}\int_0^t \frac{\dot{\gamma}(\tau)^2}{2}\mathrm{d}\tau - \frac{i}{\hbar}\int_0^t [\frac{1}{2}\gamma(\tau)A^2\gamma(\tau) + \lambda C(\gamma(\tau),\gamma(\tau),\gamma(\tau),\gamma(\tau))]\mathrm{d}\tau} \varphi(\gamma(0))\mathrm{d}\gamma \text{''},$$

$$\tag{10.31}$$

as the analytic continuation (in the parameter λ, from $\lambda < 0$ to $\lambda \geq 0$) of an infinite dimensional generalized oscillatory integral on a suitable Hilbert space. For the discussion which follows, it is convenient to modify the heuristic expression (10.31) by introducing the change of variables $\gamma(\tau) \mapsto \gamma(t-\tau)$ and obtaining:

$$\psi(t,x) =$$
$$\text{``} \int_{\gamma(0)=x} e^{\frac{i}{\hbar}\int_0^t \frac{\dot{\gamma}(\tau)^2}{2}\mathrm{d}\tau - \frac{i}{\hbar}\int_0^t [\frac{1}{2}\gamma(\tau)A^2\gamma(\tau) + \lambda C(\gamma(\tau),\gamma(\tau),\gamma(\tau),\gamma(\tau))]\mathrm{d}\tau} \varphi(\gamma(t))\mathrm{d}\gamma \text{''}.$$

$$\tag{10.32}$$

Let us consider the Hilbert space $(\mathcal{H}_t, (\ ,\ ))$ of absolutely continuous paths $\gamma : [0, t] \to \mathbb{R}^d$ with square integrable weak derivative $\int_0^t \dot{\gamma}(\tau)^2 d\tau < \infty$, fixed initial point $\gamma(0) = 0$ and inner product $(\gamma_1, \gamma_2) = \int_0^t \dot{\gamma}_1(s) \dot{\gamma}_2(s) ds$. The cylindrical Gaussian measure on $\mathcal{H}_t$ with covariance operator the identity extends to a σ-additive measure on the Wiener space $C_t = \{\omega \in C([0, t]; \mathbb{R}^d) \mid \omega(0) = 0\}$: the Wiener measure P. $(i, \mathcal{H}_t, C_t)$ is an abstract Wiener space.

Let us consider moreover the Hilbert space $\mathcal{H} = \mathbb{R}^d \times \mathcal{H}_t$, and the Banach space $\mathcal{B} = \mathbb{R}^d \times C_t$ endowed with the product measure $N(dx) \times P(d\omega)$, N being the Gaussian measure on $\mathbb{R}^d$ with covariance equal to the $d \times d$ identity matrix. $(i, \mathcal{H}, \mathcal{B})$ is an abstract Wiener space.

Let us consider two vectors $\varphi_1, \varphi_2 \in L_2(\mathbb{R}^d) \cap \mathcal{F}(\mathbb{R}^d)$. Under suitable assumptions on φ_1, φ_2, the following infinite dimensional oscillatory integral on $\mathcal{H}$:

$$
\text{``}\widetilde{\int_{\mathbb{R}^d \times \mathcal{H}_t}^{\circ}} \bar{\varphi}_1(x) e^{\frac{i}{2\hbar} \int_0^t \dot{\gamma}(\tau)^2 d\tau} e^{-\frac{i}{2\hbar} \int_0^t (\gamma(\tau)+x) A^2 (\gamma(\tau)+x) d\tau}
$$

$$
e^{-\frac{i\lambda}{\hbar} \int_0^t C(\gamma(\tau)+x, \gamma(\tau)+x, \gamma(\tau)+x, \gamma(\tau)+x) d\tau} \varphi_2(\gamma(t) + x) dx d\gamma\text{''} \quad (10.33)
$$

is well defined and can be explicitly computed in terms of an absolutely convergent integral by means of the generalized Parseval type equality (Theorem 10.7).

Let us consider the operator $L : \mathcal{H} \to \mathcal{H}$ given by:

$$
(x, \gamma) \longrightarrow (y, \eta) = L(x, \gamma),
$$

$$
y = tA^2 x + A^2 \int_0^t \gamma(\tau) d\tau, \qquad \eta(\tau) = A^2 x \left(t\tau - \frac{\tau^2}{2}\right) - \int_0^\tau \int_t^u A^2 \gamma(r) dr du
$$

$$
(10.34)
$$

and the fourth order tensor operator B:

$$
B((x_1, \gamma_1), (x_2, \gamma_2), (x_3, \gamma_3), (x_4, \gamma_4)) =
$$

$$
= \int_0^t C(\gamma_1(\tau) + x_1, \gamma_2(\tau) + x_2, \gamma_3(\tau) + x_3, \gamma_4(\tau) + x_4) d\tau. \quad (10.35)
$$

Let us consider moreover the function $f : \mathcal{H} \to \mathbb{C}$

$$
f(x, \gamma) = (2\pi i\hbar)^{d/2} e^{-\frac{i}{2\hbar}|x|^2} \bar{\varphi}_1(x) \varphi_2(\gamma(t) + x). \quad (10.36)
$$

With this notation expression (10.33) can be written in the following form:

$$
\widetilde{\int_{\mathcal{H}}^{\circ}} e^{\frac{i}{2\hbar}(|x|^2 + |\gamma|^2)} e^{-\frac{i}{2\hbar}((x,\gamma), L(x,\gamma))} e^{-\frac{i\lambda}{\hbar} B((x,\gamma),(x,\gamma),(x,\gamma),(x,\gamma))} f(x, \gamma) dx d\gamma.
$$

$$
(10.37)
$$

In the following we shall denote by A_i, $i = 1, \ldots, d$, the eigenvalues of the matrix A.

Theorem 10.8. *Let us assume that $\lambda \leq 0$, and that for each $i = 1, \ldots, d$ the following inequalities are satisfied*

$$A_i t < \frac{\pi}{2}, \qquad 1 - A_i \tan(A_i t) > 0. \tag{10.38}$$

Let $\varphi_1, \varphi_2 \in L_2(\mathbb{R}^d) \cap \mathcal{F}(\mathbb{R}^d)$. Let μ_2 be the complex bounded variation measure on $\mathbb{R}^d$ such that $\hat{\mu}_2 = \varphi_2$. Let μ_1 be the complex bounded variation measure on $\mathbb{R}^d$ such that $\hat{\mu}_1(x) = (2\pi i\hbar)^{d/2} e^{-\frac{i}{2\hbar}|x|^2} \bar{\varphi}_1(x)$. Assume in addition that the measures μ_1, μ_2 satisfy the following assumption:

$$\int_{\mathbb{R}^d} \int_{\mathbb{R}^d} e^{\frac{\hbar}{4} x A^{-1} \tan(At) x} e^{(y + \cos(At)^{-1} x)(1 - A\tan(At))^{-1}(y + \cos(At)^{-1} x)}$$

$$d|\mu_2|(x) d|\mu_1|(y) < \infty \quad (10.39)$$

Then the function $f : \mathcal{H} \to \mathbb{C}$, given by (10.36) is the Fourier transform of a bounded variation measure μ_f on $\mathcal{H}$ satisfying

$$\int_{\mathcal{H}} e^{\frac{\hbar}{4}((y,\eta),(I-L)^{-1}(y,\eta))} d|\mu_f|(y, \eta) < \infty \tag{10.40}$$

(L being given by (10.34)) and the infinite dimensional oscillatory integral (10.37) is well defined and is given by:

$$\int_{\mathbb{R}^d \times \mathcal{H}_t} \left(\int_{\mathbb{R}^d \times C_t} e^{ie^{i\pi/4}(x \cdot y + \sqrt{\hbar} n(\gamma)(\omega))} e^{\frac{1}{2\hbar} \int_0^t (\sqrt{\hbar}\omega(\tau) + x) A^2 (\sqrt{\hbar}\omega(\tau) + x) d\tau} \right.$$

$$\left. e^{i\frac{\lambda}{\hbar} \int_0^t C(\sqrt{\hbar}\omega(\tau) + x, \sqrt{\hbar}\omega(\tau) + x, \sqrt{\hbar}\omega(\tau) + x, \sqrt{\hbar}\omega(\tau) + x) d\tau} dP(\omega) \frac{e^{-\frac{|x|^2}{2\hbar}}}{(2\pi\hbar)^{d/2}} dx \right) d\mu_f(y, \gamma). \tag{10.41}$$

This is also equal to

$$(i)^{d/2} \int_{\mathbb{R}^d \times C_t} e^{i\frac{\lambda}{\hbar} \int_0^t C(\sqrt{\hbar}\omega(\tau) + x, \sqrt{\hbar}\omega(\tau) + x, \sqrt{\hbar}\omega(\tau) + x, \sqrt{\hbar}\omega(\tau) + x) d\tau}$$

$$e^{\frac{1}{2\hbar} \int_0^t (\sqrt{\hbar}\omega(\tau) + x) A^2 (\sqrt{\hbar}\omega(\tau) + x) d\tau} \bar{\varphi}(e^{i\pi/4} x) \psi_0(e^{i\pi/4} \sqrt{\hbar}\omega(t) + e^{i\pi/4} x) dP(\omega) dx. \tag{10.42}$$

Moreover the absolutely convergent integrals (10.41) and (10.42) are analytic functions of the complex variable λ if $Im(\lambda) > 0$, and continuous in $Im(\lambda) = 0$. In particular when $\lambda \geq 0$ they represent the scalar product between φ_1 and the solution of the Schrödinger equation (10.29) with Hamiltonian (10.30) and initial datum φ_2.

For detailed proofs of these results see [105], where in addition the asymptotic expansion of the oscillatory integral (10.37) in powers of the coupling constant λ is computed and its Borel summability is proved.

The results in the present section show a strong connection between two different approaches to the rigorous mathematical definition of Feynman path integrals: the analytic continuation approach and the infinite dimensional oscillatory integral approach. In fact, under suitable restrictions on the integrable function f and on the phase function, an infinite dimensional oscillatory integral on a Hilbert space $\mathcal{H}$ is exactly equal to a Gaussian integral (Theorem 10.7). As a consequence, under suitable assumptions on the initial vector φ, an infinite dimensional oscillatory integral of a function on the Hilbert space of absolutely continuous paths $\gamma : [0, t] \to \mathbb{R}^d$ with square integrable weak derivative and fixed initial point $\gamma(0) = 0$ is exactly equal to a Wiener integral of the same function after a suitable analytic continuation.

10.3 The Stationary Phase Method and the Semiclassical Expansion

One of the most fascinating features of Feynman's heuristic representation (1.13) for the solution of the Schrödinger equation is the fact that it creates a connection between the classical Lagrangian description of the physical world and the quantum one. In fact it provides a quantization method, allowing, at least heuristically, to associate to each classical Lagrangian a quantum evolution. Moreover it makes very intuitive the study of the semiclassical limit of quantum mechanics, i.e. the study of the detailed behavior of the solution of the Schrödinger equation when the Planck constant $\hbar$ is regarded as a small parameter converging to 0. In fact when $\hbar$ is small, the integrand $e^{\frac{i}{\hbar} S_t(\gamma)}$ in (1.13) is strongly oscillating and the main contribution to the integral should come (in analogy with the classical stationary phase method for finite dimensional integrals, see e.g. [123, 226, 173, 231, 230]) from those paths γ that make stationary the phase functional S_t. These, by Hamilton's least action principle, are exactly the classical orbits of the system.

Fresnel integrals of Chaps. 2–9 and infinite-dimensional oscillatory integrals in the sense of Definition 10.3 provide not only a rigorous mathematical realization for Feynman's heuristic path integral formula (1.13), but also allow the implementation of a rigorous infinite-dimensional version of the stationary phase method [87, 69] and the corresponding study of the asymptotic semiclassical expansion of the solution of the Schrödinger equation in the limit $\hbar \to 0$ [87, 69, 66, 67].

The first results can be found in [87] and were obtained in the framework of Fresnel integral approach of Chaps. 2–9. The authors consider Fresnel integrals of the form

$$I(\hbar) = \int_{\mathcal{H}}^{\sim} e^{\frac{1}{2\hbar}|x|^2} e^{-\frac{i}{\hbar} V(x)} g(x) \mathrm{d}x, \qquad (10.43)$$

where $\mathcal{H}$ is a real separable Hilbert space and V and g are in $\mathcal{F}(\mathcal{H})$, and prove, under additional regularity assumptions on V, g, that if the phase function

$\frac{1}{2}|x|^2 - V(x)$ has only non degenerate critical points, then $I(\hbar)$ is a C^∞ function of $\hbar$ and its asymptotic expansion at $\hbar = 0$ depends only on the derivatives of V and g at these critical points.

Theorem 10.9 ([87]). *Let $\mathcal{H}$ be a real separable Hilbert space, and V and g in $\mathcal{F}(\mathcal{H})$, i.e. there are bounded complex measures on $\mathcal{H}$ such that*

$$V(x) = \int_{\mathcal{H}} e^{ix\alpha} d\mu(\alpha) \qquad g(x) = \int_{\mathcal{H}} e^{ix\alpha} d\nu(\alpha).$$

Let us assume V and g C^∞, i.e. all moments of μ and ν exist. Moreover we assume $\mathcal{H} = \mathcal{H}_1 \oplus \mathcal{H}_2$ where $\dim \mathcal{H}_2 < \infty$, and if $d\mu(\beta, \gamma), d\nu(\beta, \gamma)$ are the measures on $\mathcal{H}_1 \times \mathcal{H}_2$ given by μ and ν, then there is a λ such that $\|\mu\| < \lambda^2$ and

$$\int_{\mathcal{H}} e^{\sqrt{2}\lambda|\beta|} d|\mu|(\beta, \gamma) < \infty, \qquad \int_{\mathcal{H}} e^{\sqrt{2}\lambda|\beta|} d|\nu|(\beta, \gamma) < \infty.$$

Then if the equation $dV(x) = x$ has only a finite number of solutions $x_1, \ldots, x_n$ on the support of the function g, such that none of the operators $I - d^2V(x_i)$, $i = 1, \ldots, n$, has zero as an eigenvalue, then the function

$$I(\hbar) = \int_{\mathcal{H}}^{\sim} e^{\frac{i}{2\hbar}|x|^2} e^{-\frac{i}{\hbar}V(x)} g(x) dx$$

is of the following form

$$I(\hbar) = \sum_{k=1}^{n} e^{\frac{i}{2\hbar}|x_k|^2 - V(x_k)} I_k^*(\hbar),$$

where $I_k^*(\hbar)$ $k = 1, \ldots, n$ are C^∞ functions of $\hbar$ such that

$$I_k^*(0) = e^{\frac{i\pi}{2}n_k} |\det(I - d^2V(x_k))|^{-\frac{1}{2}} g(x_k)$$

where n_k is the number of negative eigenvalues of the operator $d^2V(x_k)$ which are larger than 1.

Moreover if $V(x)$ is gentle, that is there exists a constant $\bar{\lambda} > 0$ with

$$\|\mu\| < \bar{\lambda}^2 \qquad and \qquad \int_{\mathcal{H}} e^{\sqrt{2}\bar{\lambda}|\alpha|} d|\mu|(\alpha) < \infty, \tag{10.44}$$

then the solutions of equation $dV(x) = x$ have no limit points.

Proof. If condition (10.44) is satisfied, it is possible to prove that there exists a decomposition $\mathcal{H} = \mathcal{H}_1' \oplus \mathcal{H}_2'$, with $\mathcal{H}_2 \subseteq \mathcal{H}_2'$ and $\mathcal{H}_2'$ finite dimensional such that:

$$\frac{1}{\bar{\lambda}^2} \int_{\mathcal{H}_1'} e^{\bar{\lambda}\sqrt{2}|\beta'|} d|\mu|(\beta') \leq \frac{1}{\bar{\lambda}^2} \int_{\mathcal{H}} e^{\bar{\lambda}\sqrt{2}|\beta'|} d|\mu|(\beta', \gamma) < 1. \tag{10.45}$$

So, if necessary by using the decomposition $\mathcal{H} = \mathcal{H}_1' \oplus \mathcal{H}_2'$ instead of $\mathcal{H} = \mathcal{H}_1 \oplus \mathcal{H}_2$, we may assume that, with the notation of the theorem,

$$\frac{1}{\bar{\lambda}^2} \int_{\mathcal{H}_1} e^{\bar{\lambda}\sqrt{2}|\beta|} d|\mu|(\beta) \leq \frac{1}{\bar{\lambda}^2} \int_{\mathcal{H}} e^{\bar{\lambda}\sqrt{2}|\beta|} d|\mu|(\beta,\gamma) < 1. \qquad (10.46)$$

Condition (10.46) implies that the equation

$$d_1 V(y,z) = y$$

has a unique solution $y = b(z)$ and $z \mapsto b(z)$ is a smooth mapping of $\mathcal{H}_1$ into $\mathcal{H}_2$. By using the Fubini theorem for oscillatory integrals (see Proposition 2.4), $I(\hbar)$ is equal to

$$I(\hbar) = \int_{\mathcal{H}_2}^{\sim} e^{\frac{i}{2\hbar}|z|^2} e^{\frac{i}{2\hbar}b(z)^2 - \frac{i}{\hbar}V(b(z),z)} I_2(\hbar,z) dz, \qquad (10.47)$$

with $I_2(\hbar,z) = e^{-\frac{i}{\hbar}(\frac{b(z)^2}{2} - V(b(z),z))} I_1(\hbar,z)$ and

$$I_1(\hbar,z) = \int_{\mathcal{H}_1}^{\sim} e^{\frac{i}{2\hbar}|y|^2} e^{-\frac{i}{\hbar}V(y,z)} g(y,z) dy. \qquad (10.48)$$

It is now possible to prove that the Fresnel integral $I_2(\hbar)$ on the infinite dimensional Hilbert space $\mathcal{H}_1$ is a C^∞ function of $\hbar$ on the real line and it is analytic in $\mathrm{Im}\hbar < 0$. Moreover $I_2(0) = |1 - d_1^2 V(b(z),z)|^{-1/2} g(b(z),z)$. The integral $I(\hbar)$ on $\mathcal{H}_2$ can now be studied by means of the existing theory of stationary phase for the asymptotic expansions of finite dimensional Fresnel integrals (see [87] for the more details). $\square$

In 1985 Rezende, assuming some additional regularity conditions for V and g and by considering a phase function $\frac{1}{2}(x, Bx) - V(x)$ (where B and B^{-1} are bounded symmetric operators on $\mathcal{H}$), proved the Borel summability of the asymptotic expansion in powers of the parameter $\hbar$ of the integral

$$I(\hbar) = \int_{\mathcal{H}}^{\sim} e^{\frac{i}{2\hbar}(x,Bx)} e^{-\frac{i}{\hbar}V(x)} g(x) dx.$$

(For the concept of Borel summability see, e.g. [385, 423, 130, 227]).

Theorem 10.10 ([396]). *Let $V(x) = \int_{\mathcal{H}} e^{ix\alpha} d\mu(\alpha)$ and $g(x) = \int_{\mathcal{H}} e^{ix\alpha} d\nu(\alpha)$, where μ and ν are bounded complex measures on $\mathcal{H}$ such that*

$$\int |\alpha|^n d|\mu|(\alpha) \leq Ln!/\epsilon^n, \qquad 0 < n,$$

$$\int |\alpha|^n d|\nu|(\alpha) \leq Mn!/\epsilon^n, \qquad 0 \leq n,$$

for some $L, M, \epsilon > 0$ verifying $2L\|B^{-1}\|(3 + 2\sqrt{2}) < \epsilon^2$ ($\|B^{-1}\|$ denoting the operator norm of B^{-1}).

Then there is a unique point $a \in \mathcal{H}$ such that $dV(a) = Ba$; $B^{-1}d^2V(a)$ is of trace class and its trace-norm $\| \ \|_1$ satisfies the inequality $\|B^{-1}d^2V(a)\|_1 < 1$. Let

$$I(\hbar) = \int_{\mathcal{H}} e^{\frac{i}{2\hbar}(x,Bx)} e^{-\frac{i}{\hbar}V(x)} g(x) dx.$$

Then $I(\hbar)$ is analytic in $\mathrm{Im}(\hbar) < 0$ and the function I^*, given by

$$I^*(\hbar) = I(\hbar) e^{\frac{i}{\hbar}V(a) - \frac{i}{2\hbar}(a,Ba)}, \qquad \text{for } \mathrm{Im}(\hbar) \leq 0, \hbar \neq 0,$$

is a continuous function of $\hbar$ in $\mathrm{Im}(\hbar) \leq 0$, with

$$I^*(0) = \det(I - B^{-1}d^2V(a))^{-1/2} g(a),$$

where $\det(I - B^{-1}d^2V(a))$ is the Fredholm determinant of the operator $(I - B^{-1}d^2V(a))$. Moreover one has the following asymptotic expansion and estimate

$$\left| I^*(\hbar) - \sum_{m=0}^{l-1} \hbar^m \left(-\frac{i}{2}\right)^m \sum_{n=0}^{\infty} \frac{(-2)^{-n}}{n!(m+n)!} \right.$$

$$\left. \int \cdots \int \left\{ B^{-1/2} \Big(\sum_{j=1}^{n} \alpha_j + \beta \Big) \right\}_{(\alpha,2)}^{2m+2n} \prod_{j=1}^{n} e^{i(a,\alpha_j)} d\mu(\alpha_j) e^{i(a,\beta)} d\nu(\beta) \right|$$

$$= \left| I^*(\hbar) - \det(I - B^{-1}d^2V(a))^{-1/2} \sum_{m=0}^{l-1} \hbar^m \left(-\frac{i}{2}\right)^m \sum_{n=0}^{2m} \frac{(-2)^{-n}}{n!(m+n)!} \right.$$

$$\int \cdots \int \left\{ (B - d^2V(a))^{-1/2} \Big(\sum_{j=1}^{n} \alpha_j + \beta \Big) \right\}_{(\alpha,3)}^{2m+2n}$$

$$\left. \prod_{j=1}^{n} e^{i(a,\alpha_j)} d\mu(\alpha_j) e^{i(a,\beta)} d\nu(\beta) \right|$$

$$\leq \frac{M}{(2-\sqrt{2})\sqrt{\pi}} \left(\frac{2|\hbar| \|B^{-1}\|}{\epsilon^2(6 - 4\sqrt{2})} \right)^l \left(1 - \frac{2L \|B^{-1}\|(3 + 2\sqrt{2})}{\epsilon^2} \right)^{-l-1/2} \left(l - \frac{1}{2} \right)!$$

where

$$(x_1 + \cdots + x_n + y)_{(x,m)}^{s} = \frac{s!}{(s-mn)!(m-1)!n} \int_0^t \cdots \int_0^t [(1-t_1) \cdots (1-t_n)]^{m-1}$$

$$(x_1^m, \ldots, x_n^m, (t_1 x_1 + \cdots + t_n x_n + y)^{s-mn}) dt_1 \ldots dt_n,$$

and

$$(x_1, \ldots, x_{2n}) = \frac{1}{(2n)!} \sum_{\sigma} (x_{\sigma(1)}, x_{\sigma(2)}) \cdots (x_{\sigma(2n-1)}, x_{\sigma(2n)}),$$

the summation being over all permutations σ of $\{1, \ldots, 2n\}$.
Moreover the asymptotic expansion is Borel summable and determines $I^*(\hbar)$ uniquely.

Analogous results have been proven [69] also in the framework of infinite dimensional oscillatory integrals.

Theorems 10.9 and 10.10 can be applied to the study of the asymptotic behavior of the solution of the Schrödinger equation (10.15), by using the Feynman path integral representation. The first rigorous results in this direction can be found in [87] in the framework of infinite dimensional Fresnel integrals. Analogous results concerning the Schrödinger equation with a magnetic field can be found in [69] in the framework of infinite dimensional oscillatory integrals. See also [447] for the discussion of the quasiclassical representation and a comparison with Maslov's results [38]. We summarize here the formulation of [87] (see also [67, 69]). The authors consider a particular but physically relevant form for the initial wave function $\varphi(x) = e^{\frac{i}{\hbar} f(x)} \chi(x)$, where f is real and f, χ are independent of $\hbar$. This initial data corresponds to an initial particle distribution $\rho_0(x) = |\chi|^2(x)$ and to a limiting value of the probability current $J_{\hbar=0} = f'(x)\rho_0(x)/m$, giving an initial particle flux associated to the velocity field $f'(x)/m$ (f' stands for the gradient of f) .

Theorem 10.11. ([87]) *Consider the Schrödinger equation*

$$i\hbar \frac{\partial}{\partial t} \psi = -\frac{\hbar^2}{2m} \triangle \psi + V(x)\psi$$

where the potential V is the Fourier transform of some complex measure ν such that

$$V(x) = \int_{\mathbb{R}^d} e^{ix\beta} d\nu(\beta),$$

with

$$\int_{\mathbb{R}^d} e^{|\beta|\epsilon} d|\nu|(\beta) < \infty$$

for some $\epsilon > 0$. Let the initial condition be

$$\psi(y, 0) = e^{\frac{i}{\hbar} f(y)} \chi(y)$$

with $\chi \in C_0^\infty(\mathbb{R}^d)$ and $f \in C^\infty(\mathbb{R}^d)$ and such that the Lagrange manifold $L_f \equiv (y, -\nabla f)$ intersects transversally the subset Λ_V of the phase space made of all points (y, p) such that p is the momentum at y of a classical particle that starts at time zero from x, moves under the action of V and ends at y at time t.

Then $\psi(t, x)$, given by the Feynman path integral (defined rigorously as an ∞ dimensional Fresnel integral)

$$\widetilde{\int}_{\gamma(t)=x} e^{\frac{i}{2\hbar} \int_0^t \dot{\gamma}(\tau)^2 d\tau} e^{-\frac{i}{\hbar} \int_0^t V(\gamma(\tau)) d\tau} \psi(\gamma(0), 0) d\gamma = \widetilde{\int}_{\gamma(t)=x} e^{\frac{i}{\hbar} S(\gamma)} \psi(\gamma(0), 0) d\gamma,$$

has an asymptotic expansion in powers of $\hbar$, whose leading term is the sum of the values of the function

$$\left| \det\left(\left(\frac{\partial \bar{\gamma}_k^{(j)}}{\partial y_l^{(j)}} (y^{(j)}, t) \right) \right) \right|^{-1/2} \left(e^{-\frac{i}{2}\pi m^{(j)}} e^{-\frac{i}{\hbar} S} e^{-\frac{i}{\hbar} f} \chi \right) (\bar{\gamma}^{(j)})$$

taken at the points $y^{(j)}$ such that a classical particle starting at $y^{(j)}$ at time zero with momentum $\nabla f(y^{(j)})$ is in x at time t. $S(\bar{\gamma}^{(j)})$ is the classical action along this classical path $\bar{\gamma}^{(j)}$ and $m^{(j)}(\bar{\gamma}^{(j)})$ is the Maslov index of the path $\bar{\gamma}^{(j)}$, i.e. $m^{(j)}$ is the number of zeros of $\det\left(\left(\frac{\partial \bar{\gamma}_k^{(j)}}{\partial y_l^{(j)}}(y^{(j)}, \tau)\right)\right)$ *as τ varies on the interval $(0, t)$.*

If some critical point of the phase function is degenerate, the study of the asymptotic behavior of the integral $I(\hbar)$ in (10.43) becomes more complicated. In fact we know from the case of a finite dimensional Hilbert space $\mathcal{H}$ that in this situation it is possible that the integral $I(\hbar)$, divided by the above leading term, will not tend to a limit as $\hbar \to 0$. The problem is solved in some situations by letting the functions V and g depend on an additional parameter $y \in \mathbb{R}^k$, for suitable k. In [87] it is proved that the same technique can be generalized to the infinite dimensional case.

More detailed results are presented in [69] and [67]. In [69] the authors consider phase functions which can degenerate and, under suitable assumptions, they manage to reduce the study of the degeneracy to the one of degeneracy on a finite dimensional subspace of the Hilbert space $\mathcal{H}$ and apply the existing theory for finite dimensional oscillatory integrals. In fact they assume that the phase function $\frac{1}{2}(x, Bx) - V(x)$ has the point $x_c = 0$ as a unique stationary point, which is degenerate, i.e. $Z := \mathrm{Ker}(B - \mathrm{d}^2 V)(0) \neq \{0\}$. Under suitable assumptions on B and V, they prove that Z is finite dimensional. By taking the subspace $Y = B(Z^\perp)$ and applying the Fubini theorem one has

$$I(\hbar) = \int_{\mathcal{H}}^{\widetilde{\circ}} e^{\frac{i}{2\hbar}(x,Bx)} e^{-\frac{i}{\hbar}V(x)} g(x)\mathrm{d}x =$$

$$= C_B \int_Z^{\widetilde{\circ}} e^{\frac{i}{2\hbar}(z,B_2 z)} \int_Y^{\widetilde{\circ}} e^{\frac{i}{2\hbar}(y,B_1 y)} e^{-\frac{i}{\hbar}V(y+z)} \mathrm{d}y\mathrm{d}z, \quad (10.49)$$

where B_1 and B_2 are defined by

$$B_1 y = (\pi_Y \circ B)(y), \qquad y \in Y,$$
$$B_2 z = (\pi_Z \circ B)(z), \qquad z \in Z,$$

and $C_B = (\det B)^{-1/2}(\det B_1)^{1/2}(\det B_2)^{1/2}$. By assuming that $V, g \in \mathcal{F}(\mathcal{H})$, $V = \hat{\mu}$ and $g = \hat{\nu}$, and under some growth conditions on μ and ν, one has that the phase function

$$y \mapsto \frac{1}{2}(y, B_1 y) - V(y+z)$$

of the oscillatory integral $J(z, \hbar) = \int_Y^{\widetilde{\circ}} e^{\frac{i}{2\hbar}(y,B_1 y)} e^{-\frac{i}{\hbar}V(y+z)}\mathrm{d}y$ has only one nondegenerate stationary point $a(z) \in Y$. By applying then the theory developed for the nondegenerate case one has

$$J(z, \hbar) = e^{\frac{i}{2\hbar}(a(z),B_1 a(z))} e^{-\frac{i}{\hbar}V(a(z)+z)} J^*(z, \hbar),$$

$$J^*(z, 0) = \left[\det\left(B_1 - \frac{\partial^2 V}{\partial^2 y}(a(z) + z)\right)\right]^{-1/2} g(a(z) + z).$$

As $I(\hbar) = \int_Z^{\widetilde{\circ}} e^{\varphi(z)} J^*(z,\hbar) dz$, where $\varphi(z) = \frac{i}{2\hbar}(z, B_2 z) + \frac{i}{2\hbar}(a(z), B_1 a(z)) - \frac{i}{\hbar} V(a(z)+z)$, the main contribution to the asymptotic behavior of $I(\hbar)$ comes from $J^*(z,0)$. The phase function φ has $z = 0$ as a unique degenerate critical point and, by applying the theory for the asymptotic behavior for finite dimensional oscillatory integrals [279], one has to investigate the higher derivatives of φ at 0. For example if $dim(Z) = 1$ and $\frac{\partial^3 V}{\partial^3 z}(0) \neq 0$ then

$$I(\hbar) \sim C\hbar^{-1/6}, \qquad \text{as} \quad \hbar \to 0.$$

More generally it is possible to handle other cases, taking into account the classification of different types of degeneracies (see, e.g., [69]).

A quite different situation is handled in [64], [66] and [67], where the Feynman path integral representation $I(t,\hbar)$ for the trace of the Schrödinger group $\mathrm{tr}\, e^{-\frac{i}{\hbar} Ht}$ is studied, as well as its asymptotic behavior as $\hbar \to 0$. More precisely in [67] the oscillatory integral

$$I(t,\hbar) = \int_{\mathcal{H}_{p,t}}^{\widetilde{\circ}} e^{\frac{i}{\hbar}\Phi(\gamma)} d\gamma,$$

is considered, where $\mathcal{H}_{p,t}$ is the Hilbert space of periodic functions $\gamma \in H^1(0,t;\mathbb{R}^d)$ such that $\gamma(0) = \gamma(t)$, with norm $|\gamma|^2 = \int_0^t \dot\gamma(\tau)^2 d\tau + \int_0^t \gamma(\tau)^2 d\tau$, and $\Phi(\gamma) = \frac{1}{2}\int_0^t \dot\gamma(\tau)^2 d\tau - \int_0^t V_1(\gamma(\tau)) d\tau$, $V_1(x) = \frac{1}{2} x\Omega^2 x + V_0(x)$ being the classical potential. If $V_1 : \mathbb{R}^d \to \mathbb{R}$ is of class C^2, then one proves that the functional Φ is of class C^2 and a path $\gamma \in \mathcal{H}_{p,t}$ is a stationary point for Φ if and only if γ is a solution of the Newton equation

$$\ddot\gamma(\tau) + V_1'(\gamma(\tau)) = 0 \tag{10.50}$$

satisfying the periodic conditions

$$\gamma(0) = \gamma(t), \qquad \dot\gamma(0) = \dot\gamma(t). \tag{10.51}$$

V_1 is also assumed to satisfy the following conditions:

1. V_1 has a finite number critical points $c_1, \ldots, c_s$, and each of them is non-degenerate, i.e. $\det V_1''(c_j) \neq 0$
2. $t > 0$ is such that the function γ_{c_j}, given by $\gamma_{c_j}(\tau) = c_j$, $\tau \in [0,t]$, is a non-degenerate stationary point for Φ
3. Any non-constant $t-$periodic solution γ of (10.50) and (10.51) is a "non-degenerate periodic solution", in the sense of [222], i.e. $dim\ ker\ \Phi''(\gamma) = 1$.

Under additional assumptions, the authors prove that the set M of stationary points of the phase function Φ is a disjoint union of the following form:

$$M = \{x_{c_1}, \ldots, x_{c_s}\} \cup \bigcup_{k=1}^{r} M_k,$$

where x_{c_i}, $i = 1, \ldots s$, are nondegenerate and M_k are manifolds (diffeomorphic to S^1) of degenerate stationary points, on which the phase function is constant. Under some growth conditions on V they also prove that, as $\hbar \to 0$

$$I(t, \hbar) = \sum_{j=1}^{s} e^{\frac{i}{\hbar} t V_1(c_j)} I_j^*(\hbar) + (2\pi i \hbar)^{-1/2} \left[e^{\frac{i}{\hbar} \Phi(b_k)} |M_k I_k^{**}(\hbar) + O(\hbar) \right]$$

where c_j are the points in condition 1, $b_k \in M_k$ are all noncongruent t-periodic solutions of (10.50) and (10.51) as in condition 3, $|M_k|$ is the Riemannian volume of M_k, I_j^* and I_k^{**} are C^∞ functions of $\hbar \in \mathbb{R}$ such that, in particular,

$$I_j^*(0) = \left(\det \left[2 \left[\cos \left(t \sqrt{V''(c_j)} \right) - 1 \right] \right] \right)^{-1/2},$$

$$I_k^{**}(0) = \left(\frac{\mathrm{d}}{\mathrm{d}\epsilon} \det(R_\epsilon^k(t) - I)|_{\epsilon=1} \right)^{-1/2},$$

where $R_\epsilon^k(t)$ denotes the fundamental solution of

$$\begin{cases} \ddot{x}(\tau) = -\epsilon V''(b_k(\tau)) x(\tau), & \tau > 0, \quad \epsilon > 0, \\ x(0) = x_0, \quad \dot{x}(0) = y_0 \end{cases}$$

written as a first order system of $2d$ equations for real valued functions.

Remark 10.6. The problem of corresponding asymptotic expansions in powers of $\hbar$ for the case of the Schrödinger equation with a quartic potential requires a different treatment. For the corresponding finite dimensional approximation a detailed presentation, including Borel summability, is given in [103]. The case of the Schrödinger equation itself is discussed in [76].

10.4 Alternative Approaches to Rigorous Feynman Path Integrals

10.4.1 Analytic Continuation

One of the first tools used in the rigorous mathematical realization of Feynman path integrals was analytic continuation of Gaussian Wiener integrals (see [10, 165, 166, 167, 383] and, e.g., [217, 89, 440, 441, 175, 178, 305, 306, 308] for more recent developments).

The leading idea is the extension to the complex case of the Wiener integral representation for the solution of the heat equation

$$\begin{cases} -\frac{\partial}{\partial t} u = -\frac{1}{2m} \triangle u + V(x) u \\ u(0, x) = \varphi(x) \end{cases} \tag{10.52}$$

i.e. the Feynman–Kac formula

$$u(t,x) = \int_{W_{t,x}} e^{-\int_0^t V(\sqrt{1/m}w(\tau))d\tau} \varphi(\sqrt{1/m}\,w(t))dP_{t,x}(w), \qquad (10.53)$$

where $W_{t,x} = \{w \in C(0,t; \mathbb{R}^d) : w(0) = x\}$ and $P_{t,x}$ is the Wiener measure on $W_{t,x}$. Then one introduces in (10.52) a real parameter λ_i, $i = 1, 2, 3$, related to the time t [70]

$$-\lambda_1 \hbar \frac{\partial}{\partial t} u = -\frac{1}{2m}\hbar^2 \triangle u + V(x)u$$
$$u(t,x) = \int_{W_{t,x}} e^{-\frac{1}{\lambda_1 \hbar}\int_0^t V(\sqrt{\hbar/(m\lambda_1)}w(\tau))d\tau} \varphi(\sqrt{\hbar/(m\lambda_1)}w(t))dP_{t,x}(w),$$

resp. the Planck constant $\hbar$ [217]

$$\lambda_2 \frac{\partial}{\partial t} u = \frac{1}{2m}\lambda_2^2 \triangle u + V(x)u$$
$$u(t,x) = \int_{W_{t,x}} e^{\frac{1}{\lambda_2}\int_0^t V(\sqrt{\lambda_2/m}\,w(\tau))d\tau} \varphi(\sqrt{\lambda_2/m}\,w(t))dP_{t,x}(w),$$

resp. the mass m [383]

$$\frac{\partial}{\partial t} u = \frac{1}{2\lambda_3}\triangle u - iV(x)u,$$
$$u(t,x) = \int_{W_{t,x}} e^{-i\int_0^t V(\sqrt{1/\lambda_3}w(\tau))d\tau} \varphi(\sqrt{1/\lambda_3}w(t))dP_{t,x}(w),$$

By substituting respectively $\lambda_1 = -i$, $\lambda_2 = i\hbar$, or $\lambda_3 = -im$, one gets, at least heuristically, the Schrödinger equation (with $\hbar = 1$ in the latter case) and its solution. These procedures can be made completely rigorous under suitable conditions on the potential V and the initial datum φ, see [165, 383, 167, 168, 297, 302, 447, 307, 217, 361, 377, 472, 175, 440, 70] for the details.

There are naturally many similarities, but no "automatic translation", of properties of infinite dimensional oscillatory integrals with properties of infinite dimensional probabilistic integrals (of Wiener type) [120, 303, 265], e.g. the method of stationary phase (discussed in 10.3) corresponds to the Laplace methods, see, e.g. [102, 213, 234, 163, 202, 203, 304, 326, 327, 328, 329, 230]. Also related is the study of stochastic oscillatory integrals [290, 291, 359, 360, 391, 426, 429, 433, 435, 436, 437, 438, 439, 119, 152, 254].

Note that [441, 442, 58] handle the case of the Schrödinger equation on manifolds.

10.4.2 White Noise Calculus Approach

An alternative approach to the rigorous mathematical realization of Feynman path integrals can be found in white noise calculus [275, 427, 277, 336, 348, 206, 59]. The leading idea is not radically different from the one of the Fresnel integral approach of chapters 2-9. In fact in the case of Fresnel integrals the expression

$$(2\pi i)^{-d/2} \int_{\mathbb{R}^d} e^{\frac{i}{2}(x,x)} f(x)dx, \qquad (10.54)$$

which has not a meaning as a traditional Lebesgue integral unless f is summable, is realized as a distributional pairing between $e^{\frac{i}{2}(x,x)}/(2\pi i)^{d/2}$ and the function $f \in \mathcal{F}(\mathbb{R}^d)$, by means of the Parseval type equality:

$$\int_{\mathbb{R}^d} e^{\frac{i}{2}(x,x)} f(x) dx := \int_{\mathbb{R}^d} e^{-\frac{i}{2}(x,x)} d\mu_f(x), \qquad f(x) = \int_{\mathbb{R}^d} e^{i(x,y)} d\mu_f(y).$$

In white noise calculus the pairing is realized in a different distributional setting. Indeed by manipulating the integrand in (10.54), one has

$$(2\pi i)^{-d/2} \int_{\mathbb{R}^d} e^{\frac{i}{2}(x,x)} f(x) dx$$

$$= (2\pi i)^{-d/2} \int_{\mathbb{R}^d} e^{\frac{i}{2}(x,x)+\frac{1}{2}(x,x)} f(x) e^{-\frac{1}{2}(x,x)} dx$$

$$= \frac{1}{i^{d/2}} \int_{\mathbb{R}^d} e^{\frac{i}{2}(x,x)+\frac{1}{2}(x,x)} f(x) d\mu_G(x),$$

where the latter line can be interpreted as the distributional pairing of $i^{-d/2} e^{\frac{i}{2}(x,x)+\frac{1}{2}(x,x)}$ and f not with respect to Lebesgue measure but rather with respect to the centered Gaussian measure μ_G on $\mathbb{R}^d$ with covariance the identity.

This idea can be generalized to the infinite dimensional case (following Hida [275], see also [277]). The first step is the construction of the underlying measure space, the infinite dimensional analogous of $(\mathbb{R}^d, \mu_G)$. The starting point is a real separable Hilbert space E, with inner product $\langle\,,\,\rangle$, and vector subspaces $\mathcal{E}_1 \supset \mathcal{E}_2 \supset \ldots$, each $\mathcal{E}_p$ being a Hilbert space with inner product $\langle\,,\,\rangle_p$ such that

1. $\mathcal{E} := \cap_p \mathcal{E}_p$ is dense in E and in each $\mathcal{E}_p$,
2. $|u|_q \le |u|_p$ for every $q \ge p$ and $u \in \mathcal{E}_q$,
3. for every p, the Hilbert-Schmidt norm $\|i_{qp}\|_{HS}$ of the inclusion $i_{qp} : \mathcal{E}_q \to \mathcal{E}_p$ is finite for some $q \ge p$ and $\lim_{q \to \infty} \|i_{qp}\|_{HS} = 0$.

Identifying $E := \mathcal{E}_0$ with its dual $\mathcal{E}_0^*$, there is the chain of spaces

$$\mathcal{E} := \cap_p \mathcal{E}_p \subset \cdots \subset \mathcal{E}_2 \subset \mathcal{E}_1 \subset \mathcal{E}_0 = E \simeq \mathcal{E}_0^* \subset \mathcal{E}_{-1} \subset \mathcal{E}_{-2} \cdots \subset \mathcal{E}^* := \cup_p \mathcal{E}_{-p},$$

where $\mathcal{E}_{-p} = \mathcal{E}_p^*$, $p \in \mathbb{Z}$. In a typical example one has an Hilbert-Schmidt operator K, with Hilbert-Schmidt norm $\|K\|_{HS} < 1$, $\mathcal{E}_p$ is taken as the range $Im(K^p)$ and

$$\langle u, v \rangle_p = \langle K^{-p}u, K^{-p}v \rangle. \tag{10.55}$$

According to Milnos' theorem there is a unique probability measure μ on the Borel σ-algebra (of the weak topology) on $\mathcal{E}^*$ such that for every $x \in \mathcal{E}$ the function

$$\mathcal{E}^* \to \mathbb{R} : \varphi \mapsto (\varphi, x) := \varphi(x)$$

is a mean zero Gaussian of variance $|x|_0^2$. By unitary extension, for every $x \in E$ there is a μ-almost-everywhere defined mean 0, variance $|x|_0$, Gaussian random variable $(\cdot, x)$ on $\mathcal{E}^*$, and this extends by complex linearity to complex Gaussian random variables corresponding to elements z of the complexification $E_{\mathbb{C}}$ of E.

In the following $(\mathcal{E}^*, \mu)$ will be taken as the underlying measure space for the realization of the (at this level still heuristic) expression

$$\int_{\mathcal{E}^*} e^{\frac{1}{2}(x,x)+\frac{1}{2}(x,x)} f(x) d\mu(x). \tag{10.56}$$

As a second step one has to give a meaning to (10.56) as a suitable distributional pairing and to construct the space of functionals, or of infinite dimensional distributions, to which $e^{\frac{1}{2}(x,x)+\frac{1}{2}(x,x)}$ belongs. The starting point is the space $L^2(\mathcal{E}^*, \mu)$. It is unitary equivalent by the Hermite-Ito-Segal isomorphism $\mathcal{I}$ to the symmetric Fock space $\mathcal{F}_s(E_{\mathbb{C}})$ on $E_{\mathbb{C}}$, i.e. the Hilbert space obtained by completing the symmetric tensor algebra over $E_{\mathbb{C}}$ with respect to the inner product given by $\langle\langle\sum_n u_n, \sum_m v_m\rangle\rangle_0 = \sum_n n!\langle u_n, v_n\rangle_0$, where u_n and v_n are n–tensors and $\langle\cdot, \cdot\rangle_0$ denotes the inner product on n-tensors induced by the inner product on $E_{\mathbb{C}}$. Analogously the inner products $\langle\cdot, \cdot\rangle_p$ produce inner products $\langle\langle\cdot, \cdot\rangle\rangle_p$ on $\mathcal{F}_s(E_{\mathbb{C}})$. The Hermite-Ito-Segal isomorphism $\mathcal{I} : L^2(\mathcal{E}^*, \mu) \to \mathcal{F}_s(E_{\mathbb{C}})$, specified by

$$\mathcal{I}(e^{(\cdot,z)-(z,z)_0/2}) = \mathrm{Exp}(z)$$

for every $z \in E_{\mathbb{C}}$, where $(\cdot, z) : \mathcal{E}^* \to \mathbb{C} : \varphi \mapsto \varphi(z)$ and $\mathrm{Exp}(z) = 1 + z + \frac{z^{\otimes 2}}{2!} + \frac{z^{\otimes 3}}{3!} + \cdots \in \mathcal{F}_s(E_{\mathbb{C}})$, allows to transfer the inner products $\langle\langle\cdot, \cdot\rangle\rangle_p$ from $\mathcal{F}_s(E_{\mathbb{C}})$ to $L^2(\mathcal{E}^*, \mu)$, denoted again with $\langle\langle\cdot, \cdot\rangle\rangle_p$. A *white noise distribution* over $\mathcal{E}^*$ is defined as an element of the completion $[\mathcal{E}_{-p}]$ of $L^2(\mathcal{E}^*, \mu)$ with respect to the dual norm $\langle\langle\cdot, \cdot\rangle\rangle_{-p}$, for any integer $p \geq 0$. One has the chain of Hilbert spaces

$$[\mathcal{E}] := \cap_p[\mathcal{E}_p] \subset \cdots \subset [\mathcal{E}_2] \subset [\mathcal{E}_1] \subset [\mathcal{E}_0] =$$
$$= L^2(\mathcal{E}^*, \mu) \simeq [\mathcal{E}_0^*] \subset [\mathcal{E}_{-1}] \subset [\mathcal{E}_{-2}] \cdots \subset [\mathcal{E}^*] := \cup_p[\mathcal{E}_{-p}].$$

The elements of $[\mathcal{E}]$ are taken to be test functions over $[\mathcal{E}^*]$, which is the corresponding space of distributions.

It is important to recall that a distribution $\varphi \in [\mathcal{E}^*]$ can be characterized by its S-transform $S\varphi$ (an analogue of the finite dimensional Laplace transform) which is the function on $\mathcal{E}_{\mathbb{C}}$ defined by:

$$S\varphi(z) := (\varphi, e^{(\cdot,z)-(z,z)_0/2}), \qquad z \in \mathcal{E}_{\mathbb{C}}$$

$(,)$ is the pairing between $\varphi \in [\mathcal{E}^*]$ and $e^{(\cdot,z)-(z,z)/2} \in [\mathcal{E}]$, where $(\cdot, \cdot)_0$ denotes the complex bilinear form induced on $E_{\mathbb{C}}$ by the inner product on E. In fact it is possible to prove the following characterization theorem [330], which is a generalization of Potthoff–Streit's characterization theorem [392] (see also [259]).

Theorem 10.12. *A function* $F : \mathcal{E} \to \mathbb{C}$ *is the S-transform of an element* $\varphi \in [\mathcal{E}^*]$ *if and only if satisfies the following conditions:*

1. *For all* $z_1, z_2 \in \mathcal{E}$, *the mapping* $\lambda \mapsto F(z_1 + \lambda z_2)$, *from* $\mathbb{R}$ *into* $\mathbb{C}$ *has an entire extension to* $\lambda \in \mathbb{C}$;
2. *For some continuous quadratic form* B *on* $\mathcal{E}$ *there exists constants* $C, K > 0$ *such that for all* $z \in \mathcal{E}_{\mathbb{C}}$, $\alpha \in \mathbb{C}$,

$$|F(\alpha z)| \le C \exp(K|\alpha|^2 |B(z)|).$$

One can also recover a distribution $\varphi \in [\mathcal{E}^*]$ by means of its T-transform (an analogue of the finite dimensional Fourier transform), defined by

$$T\varphi(j) := \varphi(e^{\mathrm{i}(\cdot, j)}) = e^{\frac{1}{2}(j,j)_0} S\varphi(-\mathrm{i}j).$$

In this framework it is possible to realize the Feynman path integral representation for the fundamental solution $K(t, x; 0, y)$ of the Schrödinger equation over $\mathbb{R}^d$

$$K(t, x; 0, y) = \int_{\gamma(t)=x, \gamma(0)=y} e^{\frac{\mathrm{i}}{\hbar} \int_0^t (\dot{\gamma}^2(\tau)/2) - V(\gamma(\tau)))\mathrm{d}\tau} \mathrm{d}\gamma, \qquad t > 0, x, y \in \mathbb{R}^d$$

in terms of white noise distributions on a suitable "path space" $\mathcal{E}^*$ [427, 277, 206, 336, 348]. We limit ourselves to present here some results given in [336]. Let us consider the Hilbert space $E = L_d := L^2(\mathbb{R}) \otimes \mathbb{R}^d$, the nuclear space $\mathcal{E} = S_d := S(\mathbb{R}) \otimes \mathbb{R}^d$, i.e. the space of d−dimensional Schwartz test functions, and the corresponding dual space $\mathcal{E}^* = S_d' := S'(\mathbb{R}) \otimes \mathbb{R}^d$. Let μ be the Gaussian measure on the Borel σ-algebra of S_d' identified by its characteristic function

$$\int_{S_d'} e^{\mathrm{i}\int X(\tau)f(\tau)\mathrm{d}\tau} \mathrm{d}\mu(X) = e^{-\frac{1}{2}\int f^2(\tau)\mathrm{d}\tau}, \qquad f \in S_d.$$

Heuristically the "paths" $X \in S_d'$ can be interpreted as the "velocities" (Gaussian white noises) of Brownian paths, as the d-dimensional Brownian motion is given by $B(t) = (\int_0^t X_1(\tau)\mathrm{d}\tau, \ldots, \int_0^t X_d(\tau)\mathrm{d}\tau)$, X_i being the i_{th} component of the path X. One then considers the triple

$$[S_d] \subset L^2(S_d') \subset [S_d^*]$$

and realizes the Feynman integrand as an element of $[S_d^*]$. More precisely the paths are modeled by

$$\gamma(\tau) = x - \sqrt{\hbar} \int_\tau^t X(\sigma)\mathrm{d}\sigma := x - \sqrt{\hbar}(X, 1_{(\tau, t]}),$$

(where in the sequel we shall put $\hbar = 1$ for notation simplicity) and the Feynman integrand for the free particle is realized as the distribution

$$I_0(x, t; y, 0) = N e^{\frac{\mathrm{i}+1}{2} \int_0^t X(\tau)^2 \mathrm{d}\tau} \delta(\gamma(0) - y),$$

(where N stands for normalization and $\delta(\gamma(0) - y)$ fixes the initial point of the path), defined rigorously by its T-transform:

$$TI_0(f) = \frac{1}{(2\pi i t)^{d/2}} e^{-\frac{1}{2}\int_{\mathbb{R}} f(\tau)^2 d\tau - \frac{1}{(2it)}(\int_{\mathbb{R}} f(\tau)d\tau + x - y)^2}. \tag{10.57}$$

It is interesting to note that the values of $TI_0(f)$ for $f = 0$ as well as for a generic $f \in S_d$ have a direct physical meaning, which becomes more clear by using the following heuristic notation and a formal integration by parts

$$TI_0(f) = \int_{S'_d} e^{i\int_{\mathbb{R}} \dot\gamma(\tau)f(\tau)d\tau} N e^{\frac{i+1}{2}\int_0^t \dot\gamma(\tau)^2 d\tau} \delta(\gamma(0) - y)d\mu$$

$$= e^{-\frac{1}{2}\int_{[0,t]^c} f^2(\tau)d\tau} e^{ixf(t)-iyf(0)} \int N e^{\frac{i+1}{2}\int_0^t \dot\gamma(\tau)^2 d\tau} e^{-i\int_0^t \gamma(\tau)\dot f(\tau)d\tau} \delta(\gamma(0) - y)d\mu. \tag{10.58}$$

In fact $TI_0(f)$ gives the fundamental solution $K^f(t, x; 0, y)$ of the Schrödinger equation with time dependent linear potential $V(t, x) = \dot f(t)x$

$$(i\frac{\partial}{\partial t} + \frac{1}{2}\triangle - \dot f(t)x)K^f(t, x; 0, y) = 0, \qquad \lim_{t \to 0} K^f(t, x; 0, y) = \delta(x - y)$$

multiplied by the factor $e^{-\frac{1}{2}\int_{[0,t]^c} f^2(\tau)d\tau} e^{ixf(t)-iyf(0)}$, as one can easily verify by direct computation of expression (10.57). The same technique allows one to handle more general potentials, such as those which are Laplace transform of bounded measures (see [336] for a detailed exposition), a corresponding result has been obtained by different methods in [73] and [92]. The white noise approach has been also successfully employed in the rigorous construction of the Chern–Simons functional integral in topological field theory (see Sect. 10.5.5). For other applications of the white noise approach to Feynman path integrals see, e.g. [147, 148, 150, 176, 258, 260, 261, 414, 426].

10.4.3 The Sequential Approach

An alternative approach to the rigorous mathematical definition of Feynman path integrals which is very close to Feynman's original derivation of (1.13) is the "sequential approach". The starting point is the Lie–Kato–Trotter product formula (which is also discussed in Chap. 1, see equations (1.7)-(1.11)), which allows to write the unitary evolution operator $e^{-\frac{i}{\hbar}tH}$, whose generator is the Hamiltonian operator $H = H_0 + V$, $H_0 = -\frac{\hbar^2}{2m}\triangle$, in terms of the following strong operator limit:

$$e^{-\frac{i}{\hbar}tH} = s - \lim_{n \to \infty} \left(e^{-\frac{i}{\hbar}\frac{t}{n}V} e^{-\frac{i}{\hbar}\frac{t}{n}H_0}\right)^n. \tag{10.59}$$

By taking an vector φ in Schwartz space $S(\mathbb{R}^n)$ and by substituting into (10.59) the Green function of the unitary operator $e^{-\frac{i}{\hbar}tH_0}$:

$$e^{-\frac{i}{\hbar}tH_0}\varphi(x) = \left(2\pi i\frac{\hbar}{m}t\right)^{-\frac{d}{2}} \int e^{im\frac{(x-y)^2}{2\hbar t}}\varphi(y)dy, \qquad (10.60)$$

one gets the following expression

$$e^{-\frac{i}{\hbar}tH}\varphi(x)$$

$$= s - \lim_{n\to\infty}\left(2\pi i\frac{\hbar}{m}\frac{t}{n}\right)^{-\frac{dn}{2}}\int_{\mathbb{R}^{nd}} e^{-\frac{i}{\hbar}\sum_{j=1}^{n}\left[\frac{m}{2}\frac{(x_j-x_{j-1})^2}{\left(\frac{t}{n}\right)^2}-V(x_j)\right]\frac{t}{n}}\varphi(x_0)$$

$$dx_0\ldots dx_{n-1} \quad (10.61)$$

where $x_n = x$ and the exponent in the integrand can be recognized as the classical action functional:

$$S_t(x_n,\ldots,x_0) = \sum_{j=1}^{n}\left[\frac{m}{2}\frac{(x_j-x_{j-1})^2}{\left(\frac{t}{n}\right)^2} - V(x_j)\right]\frac{t}{n}.$$

In particular the term $\frac{(x_j-x_{j-1})}{\left(\frac{t}{n}\right)}$ is the (constant) velocity of the path connecting the points x_{j-1} and x_j in the time interval $\frac{t}{n}$.

Equation (10.61) is a special case of the semigroup product formula

$$s - \lim_{n\to\infty}(F(t/n))^n = \exp(tF'(0)) \qquad (10.62)$$

where $t \mapsto F(t)$ is a strongly continuous mapping form the reals (or non-negative reals) into the space of bounded linear operators on an Hilbert space $\mathcal{H}$, while $F'(0)$ has to be interpreted as some operator extension of the strong limit $s - \lim_{t\to 0} t^{-1}(F(t) - I)$. In particular if A, B are self-adjoint operators in $\mathcal{H}$ and $F(t) = e^{itA}e^{itB}$, one gets formally the Trotter product formula

$$s - \lim_{n\to\infty}(e^{itA/n}e^{itB/n})^n = e^{it(A+B)} \qquad (10.63)$$

(where the sum $A + B$ has to be suitably interpreted). Precursors in this approach are the works of Yu. Daleckii and others, e.g. [194, 195] (see also ref. [10]).

Nelson in 1964 [383] proved (10.63) in connection with the rigorous mathematical definition of Feynman path integrals, under the assumption that the potential V belongs to the class considered by Kato (see [10(6)]). Some time later, in 1972, C.N. Friedman [11] studied (10.62) in connection with continuous quantum observation. Feynman himself in [1] considered particular "ideal" quantum measurements of position, made to determine whether or not the trajectory of a particle lies in a certain space–time region. By substituting in (10.62) for $F(t)$ the operator $EP(t)E$, where $P(t)$ is a contraction semigroup in the Hilbert space $\mathcal{H}$ and E is an orthogonal projection, and letting $P(t) = e^{-itH_0}$ and $E : L_2(\mathbb{R}^d) \to L_2(\mathbb{R}^d)$ be the orthogonal projection given by multiplication by the characteristic function of a suitable region $\mathcal{R}$ of $\mathbb{R}^d$, the limit

$$\lim_{n \to \infty} \|(E e^{-it H_0/n} E)^n \varphi\|^2$$

(if it exists) should give the probability that a continual observation during the time interval $[0, t]$ yields the result that the particle, whose initial state is the vector $\varphi \in L_2(\mathbb{R}^d)$, lies constantly in the region $\mathcal{R}$.

The leading idea of the "sequential approach" is the definition of the Feynman path integral by means of a limiting procedure like (10.61), without using the Trotter product formula. More precisely for any $n \in \mathbb{N}$ one considers a partition of the interval $[0, t]$ into n subintervals $t_0 = 0 < t_1 < \ldots < t_j < \ldots < t_n = t$ and for each $j = 0 \ldots n$ a point $x_j \in \mathbb{R}^d$. The path γ in the heuristic expression (1.13) is then approximated by a broken line path, passing, for each $j = 0, \ldots, n$, from the point x_j at time t_j. There are two approaches to this "time slicing approximation": one can connect the point x_j at time t_j with the point x_{j+1} at time t_{j+1} by means of a straight line path [449, 335, 244], i.e. $\gamma(\tau) = x_j + \frac{x_{j+1} - x_j}{t_{j+1} - t_j} (\tau - t_j)$, $\tau \in [t_j, t_{j+1}]$, or by means of a classical path [239], i.e. the (unique for suitable V and if $|t_{j+1} - t_j|$ is sufficiently small) solution of the classical equation of motion

$$\begin{cases} m\ddot{\gamma}(\tau) = -\nabla V(\tau, \gamma(\tau)) \\ \gamma(t_j) = x_j, \\ \gamma(t_{j+1}) = x_{j+1} \end{cases}$$

An heuristic expression like

$$\int e^{\frac{i}{\hbar} S_t(\gamma)} f(\gamma) d\gamma$$

is then realized as the limit of the time slicing approximation for suitable functional f on the path space. Indeed denoting by γ_n the broken line path (straight resp. piecewise classical) associated to the partition $t_0 = 0 < t_1 < \ldots < t_j < \ldots < t_n = t$, and by $\triangle_n$ the amplitude of each time subinterval, i.e. $\triangle_n = |t_{j+1} - t_j|$, one defines

$$F(f) \equiv \int e^{\frac{i}{\hbar} S_t(\gamma)} f(\gamma) d\gamma := \lim_{\triangle_n \to 0} \prod_{j=1}^{n} \left(\frac{1}{2\pi i \hbar t_j} \right)^{d/2} \int_{\mathbb{R}^{nd}} e^{\frac{i}{\hbar} S_t(\gamma_n)} f(\gamma_n) \prod_{j=1}^{n} dx_j,$$

$$(10.64)$$

whenever the limit exists. The integrals on the r.h.s. do not converge absolutely and are meant as (finite dimensional) oscillatory integrals.

Fujiwara in the case of approximation with piecewise classical paths and Fujiwara and Kumano-go in the case of broken line paths prove the existence of the limit (10.64) for a suitable class of functionals f. They assume that the potential $V(t, x)$ is a real valued function of $(t, x) \in \mathbb{R} \times \mathbb{R}^d$ and for any multi-index α, $\partial_x^\alpha V(t,x)$ is continuous in $\mathbb{R} \times \mathbb{R}^d$. Moreover they assume that for any integer $k \geq 2$ there exists a positive constant A_k such that

$$|\partial_x^\alpha V(t,x)| \leq A_k, \qquad |\alpha| = k,$$

(this excludes polynomial behaviour at infinity). The so defined functional F has some important properties. Integration by parts and Taylor expansion formula with respect to functional differentiation hold. F is invariant under orthogonal transformations and transforms naturally under translations. Moreover the fundamental theorem of calculus holds for F and a semiclassical approximation has been developed. Moreover it is possible to interchange the order of integration with Riemann–Stieltjes integrals and to interchange the operation of taking the integral and the one of taking a limit [244].

It is interesting to note that in the particular case where $f \in \mathcal{F}(\mathcal{H})$, $F(f)$ coincides with the infinite dimensional oscillatory integral $\mathcal{F}^\hbar(f)$ (Definition 10.3). For other sequential approaches see, e.g., [156, 177, 180, 218, 156, 284, 285, 286, 289, 443, 444, 445, 247].

10.4.4 The Approach via Poisson Processes

An alternative approach to the rigorous mathematical definitions of Feynman path integrals is based on Poisson measures. It was originally proposed by Maslov and Chebotarev [365, 368, 369, 370, 367, 332] and further developed by Blanchard, Combe, Høegh-Krohn, Rodriguez, Sirugue, Sirugue-Collin, [157, 183, 184, 185, 189], and, recently, by Kolokoltsov [324].

The potentials V which can be handled by this method are those belonging to the Fresnel class $\mathcal{F}(\mathcal{H})$. The approach is based on the fact that given a function $V : \mathbb{R}^d \to \mathbb{C}$ which is the Fourier transform of a finite complex Borel measure μ_v on $\mathbb{R}^d$,

$$V(x) = \int_{\mathbb{R}^d} e^{ikx} d\mu_v(k), \qquad x \in \mathbb{R}^k,$$

under suitable assumptions it is possible to construct a probabilistic representation of the solution of the Schrödinger equation in momentum representation:

$$\begin{cases} \frac{\partial}{\partial t}\tilde{\psi}(p) = -\frac{i}{2}p^2\tilde{\psi}(p) - iV(-i\nabla_p)\tilde{\psi}(p) \\ \tilde{\psi}(0,p) = \tilde{\varphi}(p) \end{cases} \tag{10.65}$$

In fact for any $\mu_v \in \mathcal{F}(\mathbb{R}^d)$ there exist a positive finite measure ν and a complex-valued measurable function f such that $\mu_v(dk) = f(k)\nu(dk)$. Without loss of generality it is possible to assume that $\nu(\{0\}) = 0$, so that ν is a finite Lévy measure. Let us consider a Poisson process having Lévy measure ν (see, e.g. [393, 121, 122] for these concepts). This process has almost surely piecewise constant paths. More precisely a typical path P on the time interval $[0,t]$ is defined by a finite number of independent random jumps $\delta_1, ..., \delta_n$, distributed according to the probability measure ν/λ_ν, with $\lambda_\nu = \nu(\mathbb{R}^d)$, occurring at random times $\tau_1, ... \tau_n$, distributed according to a Poisson measure with intensity λ_ν. Under the assumption that $\tilde{\varphi}(p)$ is a bounded continuous function, it is possible to prove that the solution of the Cauchy problem (10.65) can be represented by the following path integral:

$$\tilde{\psi}(t,p) = e^{t\lambda_\nu} \mathbb{E}_p^{[0,t]} [e^{-\frac{i}{2}\sum_{j=0}^n (P_j,P_j)(\tau_{j+1}-\tau_j)} \prod_{j=1}^n (-if(\delta_j)) \tilde{\varphi}(P(t))],$$

where the expectation $\mathbb{E}_p^{[0,t]}$ is taken with respect to the measure associated to the Poisson process and the sample path $P(\cdot)$ is given by

$$P(\tau) = \begin{cases} P_0 = p, & 0 \leq \tau < \tau_1 \\ P_1 = p + \delta_1, & \tau_1 \leq \tau < \tau_2 \\ ... \\ P_n = p + \delta_1 + \delta_2 + ... + \delta_n, & \tau_n \leq \tau \leq t \end{cases} \qquad (10.66)$$

The present approach has also been successfully applied to the study of the Klein–Gordon equation [183, 185], to Fermi systems [184], and to the solution of the Dirac equation [332]. We refer to the above cited bibliography for a more detailed discussion.

10.5 Recent Applications

The examples we are going to describe show that infinite dimensional oscillatory integrals are a flexible tool and can provide rigorous mathematical realizations for a large class of Feynman path integral representations.

10.5.1 The Schrödinger Equation with Magnetic Fields

In [69, 70] the Feynman path integral representation for a Schrödinger equation with magnetic field is studied:

$$\begin{cases} i\hbar \frac{\partial}{\partial t}\psi = \frac{1}{2}(-i\hbar\nabla + a(x))^2\psi + V_0(x)\psi \\ \psi(0,x) = \varphi(x), & x \in \mathbb{R}^d, t > 0 \end{cases} \qquad (10.67)$$

where $a(x)$ and $V_0(x)$ are a vector and a scalar potential respectively. Let us assume that $a(x) = Cx$, where C is an anti-self-adjoint linear operator in $\mathbb{R}^d$ and $V_0, \varphi \in \mathcal{F}(\mathbb{R}^d)$. Equation (10.67) describes the time evolution of a charged quantum particle (with unitary charge and mass) in a constant magnetic field $\vec{B} = \text{rot}(Cx)$ and in a scalar potential V_0.

Let us consider the Hilbert space $\mathcal{H}_t$ of absolutely continuous functions $\gamma : [0,t] \to \mathbb{R}^d$, such that $\gamma(0) = 0$ and $\int_0^t \dot{\gamma}(\tau)^2 d\tau < \infty$, with scalar product $(\gamma_1, \gamma_2) = \int_0^t \dot{\gamma}_1(\tau)\dot{\gamma}_2(\tau)d\tau$. Let $L : \mathcal{H}_t \to \mathcal{H}_t$ be the operator on $\mathcal{H}_t$ defined by

$$(L\gamma_1, \gamma_2) = \int_0^t (C\gamma_1(\tau)\dot{\gamma}_2(\tau) + C\gamma_2(\tau)\dot{\gamma}_1(\tau))\mathrm{d}\tau.$$

It is possible to prove (see [70] for details) that L is symmetric and belongs to the Hilbert-Schmidt class $T_2(\mathcal{H}_t)$. The Feynman path integral representation for the solution of the Schrödinger equation (10.67)

$$\psi(t,x) = \widetilde{\int_{\gamma(t)=x}} e^{\frac{i}{2\hbar}\int_0^t \dot{\gamma}(\tau)^2\mathrm{d}\tau - \frac{i}{\hbar}\int_0^t C(\gamma(\tau)+x)\dot{\gamma}(\tau)\mathrm{d}\tau - \frac{i}{\hbar}\int_0^t V_0(\gamma(\tau)+x)\mathrm{d}\tau} \varphi(\gamma(0)+x)\mathrm{d}\gamma$$

can be rigorously mathematically realized as a class-2 normalized integral over the Hilbert space $\mathcal{H}_t$ (for the definition of class-p normalized integrals see Sect. 10.1). More precisely the following holds:

Theorem 10.13. *Under the above assumptions and if* $\sin(t\sqrt{C^*C}) \neq 0$, *the solution of the Schrödinger equation* (10.67) *is given by the following class 2 normalized integral:*

$$\psi(t,x) = \widetilde{\int_{\mathcal{H}_t}^2} e^{\frac{i}{2\hbar}\int_0^t \dot{\gamma}(\tau)^2\mathrm{d}\tau + \frac{i}{2\hbar}(L\gamma,\gamma)} e^{-\frac{i}{\hbar}\int_0^t V_0(\gamma(\tau)+x)\mathrm{d}\tau} \varphi(\gamma(t)+x)\mathrm{d}\gamma$$

For the proof see [70].

10.5.2 The Schrödinger Equation with Time Dependent Potentials

The aim of the present section is to provide a rigorous mathematical realization for the Feynman path integral representation for the solution of Schrödinger equation where the potential is explicitly time dependent:

$$\begin{cases} i\hbar\frac{\partial}{\partial t}\psi = (-\frac{\hbar^2}{2m}\triangle + V(t,x))\psi \\ \psi(0,x) = \varphi(x) \end{cases} \tag{10.68}$$

First of all we consider a linearly forced harmonic oscillator, i.e. let us assume that the potential V is of the type "quadratic plus linear" and that the linear part depends explicitly on time:

$$V(t,x) = \frac{1}{2}x\Omega^2 x + f(t)\cdot x, \quad x \in \mathbb{R}^d \tag{10.69}$$

where Ω is a positive symmetric constant $d \times d$ matrix with eigenvalues Ω_j, $j = 1\ldots d$, and $f : I \subset \mathbb{R} \to \mathbb{R}^d$ is a continuous function. This potential is particularly interesting from a physical point of view as it is used in simple models for a large class of processes, as the vibration-relaxation of a diatomic molecule in gas kinetics and the interaction of a particle with the field oscillators in quantum electrodynamics. Feynman calculated heuristically the Green function for (10.69) in his famous paper on the path integral formulation of quantum mechanics [232]. Under suitable assumptions on the initial datum φ it is possible to prove that Feynman's heuristic formula

$$\psi(t,x) = \int_{\gamma(t)=x} e^{\frac{i}{2\hbar}\int_0^t |\dot\gamma(\tau)|^2 d\tau - \frac{i}{2\hbar}\int_0^t \gamma(\tau)\Omega^2\gamma(\tau)d\tau - \frac{i}{\hbar}\int_0^t f(\tau)\cdot\gamma(\tau)d\tau} \psi_0(\gamma(0))d\gamma$$

(10.70)

can be realized as an infinite dimensional oscillatory integral on the Hilbert space $\mathcal{H}_t$ of absolutely continuous paths $\gamma : [0,t] \to \mathbb{R}^d$, such that $\gamma(t) = 0$, and square integrable weak derivative $\int_0^t |\dot\gamma(\tau)|^2 d\tau < \infty$, endowed with the inner product $(\gamma_1,\gamma_2) = \int_0^t \dot\gamma_1(\tau)\cdot\gamma_2(\tau)d\tau$. Let $L : \mathcal{H}_t \to \mathcal{H}_t$ be the trace class symmetric operator on $\mathcal{H}_t$ given by:

$$(L\gamma)(\tau) = \int_\tau^t \int_0^{\tau'} (\Omega^2\gamma)(\tau'')d\tau''d\tau', \qquad \gamma \in \mathcal{H}_t.$$

One can easily verify that if $t \neq (n+1/2)\pi/\Omega_j$, $n \in \mathbb{Z}$ and Ω_j any eigenvalue of Ω, then $(I - L)$ is invertible and

$$\det(I - L) = \det(\cos(\Omega t)).$$

Let moreover $w, v \in \mathcal{H}_t$ be defined by

$$w(\tau) \equiv \frac{\Omega^2 x}{2\hbar}(\tau^2 - t^2), \qquad v(\tau) \equiv \frac{1}{\hbar}\int_t^\tau \int_0^{\tau'} f(\tau'')d\tau''d\tau'. \qquad (10.71)$$

Then the heuristic Feynman path integral representation (10.70) can be realized as the following infinite dimensional oscillatory integral on $\mathcal{H}_t$:

$$\psi(t,x) = \int_{\gamma(t)=0} e^{\frac{i}{2\hbar}\int_0^t |\dot\gamma(s)|^2 ds - \frac{i}{2\hbar}\int_0^t (\gamma(s)+x)\Omega^2(\gamma(s)+x)ds}$$
$$e^{-\frac{i}{\hbar}\int_0^t f(s)\cdot(\gamma(s)+x)ds}\varphi(\gamma(0)+x)d\gamma = e^{-i\frac{t}{2\hbar}x\Omega^2 x}e^{-i\frac{x}{2\hbar}\cdot\int_0^t f(s)ds}$$
$$\widetilde{\int_{\mathcal{H}_t}^\circ} e^{\frac{i}{2\hbar}\langle\gamma,(I-L)\gamma\rangle}e^{i\langle v,\gamma\rangle}e^{i\langle w,\gamma\rangle}\varphi(\gamma(0)+x)d\gamma \qquad (10.72)$$

(where $\widetilde{\int_{\mathcal{H}_t}^\circ}$ is a rigorous functional in the sense of Definition 10.3). Indeed let $\varphi \in \mathcal{F}(\mathbb{R}^d)$, then one can easily see that the functional $\gamma \mapsto e^{i\langle v,\gamma\rangle}e^{i\langle w,\gamma\rangle}\psi_0(\gamma(0) + x)$ belongs to $\mathcal{F}(\mathcal{H}_t)$ and the infinite dimensional oscillatory integral (10.72) on $\mathcal{H}_t$ can be explicitly computed by means of the Parseval-type equality (10.6). If moreover $\varphi \in \mathcal{S}(\mathbb{R}^d)$, one can proceed further and compute explicitly the Green function $G(0,t,x,y)$:

$$\psi(t,x) = \int_{\mathbb{R}^d} G(0,t,x,y)\varphi(y)dy,$$

where

$$G(0,t,x,y) = (2\pi i\hbar)^{-d/2}\sqrt{\det\left(\frac{\Omega}{\sin(\Omega t)}\right)}e^{\frac{i\Omega\sin(\Omega t)^{-1}}{2\hbar}(x\cos(\Omega t)x+y\cos(\Omega t)y-2xy)}$$

$$e^{-\frac{i}{\hbar}x\sin(\Omega s)^{-1}\int_0^t\sin(\Omega s)f(s)ds-\frac{i}{\hbar}y(\int_0^t\cos(\Omega s)f(s)ds-\cos(\Omega t)\sin(\Omega t)^{-1}\int_0^t\sin(\Omega s)f(s)ds)}$$

$$e^{\frac{i}{\hbar}\Omega^{-1}(\frac{1}{2}\cos(\Omega t)\sin(\Omega t)^{-1}(\int_0^t\sin(\Omega s)f(s)ds)^2-\int_0^t\sin(\Omega s)f(s)ds\int_0^t\cos(\Omega s)f(s)ds)}$$

$$e^{\frac{i}{\hbar}\Omega^{-1}\int_0^t\cos(\Omega s)f(s)\int_s^t\sin(\Omega s')f(s')ds'ds} \qquad (10.73)$$

where $t \neq (n+1/2)\pi/\Omega_j$ (see [107] for more details).

An analogous result can be obtained for the Schrödinger equation with an harmonic-oscillator potential with a time-dependent frequency:

$$V(t,x) = \frac{1}{2}x\Omega^2(t)x, \qquad x \in \mathbb{R}^d \qquad (10.74)$$

where $\Omega : [0,t] \rightarrow L(\mathbb{R}^d,\mathbb{R}^d)$ is a continuous map from the time interval $[0,t]$ to the space of symmetric positive $d \times d$ matrices. This problem has been analyzed by several authors (see for instance [390, 314] and references therein) as an approximate description for the vibration of complex physical systems, as well as an exact model for some physical phenomena, as the motion of an ion in a Paul trap, the quantum mechanical description of highly cooled ions, the emergence of non classical optical states of light owing to a time-dependent dielectric constant, or even in cosmology for the study of a three-dimensional isotropic harmonic oscillator in a spatially flat universe, with metric given by $g_{ij} = R(t)\delta_{ij}$, with $R(t)$ being the scale factor at time t.

If $d = 1$, it is possible to solve the Schrödinger equation with potential (10.74) (and also the corresponding classical equation of motion) by adopting a suitable transformation of the time and space variables which allows to map the solution of the time-independent harmonic oscillator to the solution of the time-dependent one (see [397, 398, 399] and references therein). Indeed by considering the classical equation of motion for the time-dependent harmonic oscillator (10.74)

$$\ddot{u}(s) + \Omega^2(s)u(s) = 0, \qquad (10.75)$$

given two independent solutions u_1 and u_2 of (10.75) such that $u_1(0) = \dot{u}_2(0) = 0$ and $u_2(0) = \dot{u}_1(0) = 1$, it is easy to prove that the function $\xi := u_1^2 + u_2^2$ is strictly positive $\xi(s) > 0$ $\forall s$ and it satisfies the following differential equation:

$$2\xi\ddot{\xi} - \dot{\xi}^2 + 4\xi^2 - 4 = 0.$$

Moreover the function $\eta : [0,\infty] \rightarrow \mathbb{R}$

$$\eta(s) = \int_0^s \xi(\tau)^{-1}d\tau$$

is well defined and strictly increasing. One verifies that

$$u(s) = \xi(s)^{1/2}(A\cos(\eta(s)) + B\sin(\eta(s))) \tag{10.76}$$

is the general solution of the classical equation of motion (10.75). In other words by rescaling the time variable $s \mapsto \eta(s)$ and the space variable $x \mapsto \xi^{-1/2}x$ it is possible to map the solution of the equation of motion for the time-independent harmonic oscillator $\ddot{u}(s) + u(s) = 0$ into the solution of (10.75). In fact it is possible to find (see, for instance, [309] for more details) a general canonical transformation $(x, p, t) \mapsto (X, P, \tau)$, given by

$$\begin{cases} X = \xi(t)^{-1/2}x \\ \frac{d\tau(t)}{dt} = \xi(t)^{-1} \\ P = \frac{dX}{d\tau} = (\xi^{1/2}\dot{x} - \frac{1}{2}\xi^{-1/2}\dot{\xi}x) \end{cases} \tag{10.77}$$

and the Hamiltonian is given by $H(X, P, \tau) = \frac{1}{2}(P^2 + X^2)$, while the generating function of the transformation $(x, p, t) \mapsto (X, P, \tau)$ is given by $F(x, P, t) = \xi(t)^{-1/2}xP + \frac{\xi(t)^{-1}\dot{\xi}}{4}x^2$ and the transformation is given more explicitly as

$$\begin{cases} p = \frac{\partial}{\partial x}F(x, P, t) \\ X = \frac{\partial}{\partial P}F(x, P, t) \\ H(X, P; \tau)\dot{\tau} = H(x, p; t) + \frac{\partial}{\partial t}F(x, P, t) \end{cases} \tag{10.78}$$

A similar result holds also in the quantum case. In fact by considering the Schrödinger equations for the time-independent and time-dependent harmonic oscillator respectively,

$$(i\hbar\frac{\partial}{\partial t} + \frac{\hbar^2}{2}\triangle - \frac{1}{2}x^2)\psi_{TI}(t, x) = 0, \tag{10.79}$$

$$(i\hbar\frac{\partial}{\partial t} + \frac{\hbar^2}{2}\triangle - \frac{1}{2}\Omega^2(t)x^2)\psi_{TD}(t, x) = 0, \tag{10.80}$$

it is possible to prove [397] the following relation between the two solutions ψ_{TD} and ψ_{TI}

$$\psi_{TI} = \xi(t)^{-1/4}\exp[i\dot{\xi}(t)x^2/4\hbar\xi(t)]\psi_{TD}(\eta(t), \xi(t)^{-1/2}x). \tag{10.81}$$

In an analogous way it is possible to prove that, denoting by $K_{TI}(t, 0; x, y)$ and $K_{TD}(t, 0; x, y)$ the Green functions for the Schrödinger equations (10.79) and (10.80) respectively, the following holds:

$$K_{TD}(t, 0; x, y) = \xi(t)^{-1/4}\exp[i\dot{\xi}(t)x^2/4\hbar\xi(t)]K_{TI}(\eta(t), 0; \xi(t)^{-1/2}x, y). \tag{10.82}$$

It is interesting to note that the "correction term" $\xi(t)^{-1/4}\exp[i\dot{\xi}(t)x^2/4\hbar\xi(t)]$ in equations (10.81) and (10.82) can be interpreted in terms of the classical canonical transformation (10.78) (see [309] for more details).

Infinite dimensional oscillatory integrals provide a tool for the derivation of (10.81) and (10.82). Indeed let us consider the following linear operator $L : \mathcal{H}_t \to \mathcal{H}_t$:

$$(L\gamma)(\tau) = -\int_t^\tau \int_0^r \Omega^2(u)\gamma(u)dudr, \qquad \gamma \in \mathcal{H}_t.$$

One can easily verify that L is self-adjoint, trace class and positive, moreover, by using (10.76), it is possible to prove that, if $t \neq \eta^{-1}(\pi/2 + n\pi)$, $n \in \mathbb{N}$, the operator $I - L$ is invertible, its inverse can be explicitly computed and its Fredholm determinant is given by $\det(I - L) = \xi(t)^{1/2}\cos(\eta(t))$ (see [107] for more details). Let us consider the vectors $G_0, u \in \mathcal{H}_t$ given by

$$G_0(\tau) = t - \tau, \qquad u(\tau) = \frac{x}{\hbar}\int_t^\tau \int_0^u \Omega^2(r)drdu, \ x \in \mathbb{R}.$$

With the notations introduced so far and by assuming that the initial vector φ belongs to $\mathcal{F}(\mathbb{R})$, so that $\varphi = \hat{\mu}_0$, the heuristic Feynman path integral representation for the solution of the Schrödinger equation with the time-dependent potential (10.74)

$$\psi(t, x) = \ \text{``} \int_{\{\gamma|\gamma(t)=0\}} e^{\frac{i}{2\hbar}\int_0^t \dot{\gamma}^2(\tau)d\tau - \frac{i}{2\hbar}\int_0^t \Omega^2(\tau)(\gamma(\tau)+x)^2d\tau}\varphi(\gamma(0)+x)d\gamma \ \text{''}$$

can be rigorously realized as the infinite dimensional oscillatory integral associated with the Cameron-Martin space $\mathcal{H}_t$:

$$\psi(t, x) = e^{\frac{ix^2}{2\hbar}\int_0^t \Omega^2(\tau)d\tau}\widetilde{\int_{\mathcal{H}_t}}^{\circ} e^{\frac{i}{2\hbar}(\gamma,(I-L)\gamma)}e^{i(u,\gamma)}\varphi(\gamma(0) + x)d\gamma,$$

that can be computed by means of Parseval-type equality (see Theorem 10.1). Moreover if $\varphi \in \mathcal{S}(\mathbb{R})$, one can proceed further and compute the Green function $K_{TD}(t, 0; x, y)$:

$$K_{TD}(t, 0; x, y) = \xi(t)^{-1/4}e^{\frac{ix^2}{4\hbar}\xi(t)^{-1}\dot{\xi}(t)}\frac{e^{\frac{i}{2\hbar}(\frac{\cos(\eta(t))}{\sin(\eta(t))})(\xi(t)^{-1}x^2+y^2) - \frac{2\xi(t)^{-1/2}xy}{\sin(\eta(t))}}}{(2\pi i\hbar\sin(\eta(t)))^{1/2}}.$$

By recalling the well known formula for the Green function $K_{TI}(t, 0; x, y)$ of the Schrödinger equation with a time-independent harmonic oscillator Hamiltonian (see, e.g., [413]):

$$K_{TI}(t, 0; x, y) = \frac{e^{\frac{i}{2\hbar}(\frac{\cos(t)}{\sin(t)}(x^2+y^2) - \frac{2xy}{\sin(t)})}}{(2\pi i\hbar\sin(t))^{1/2}}$$

one can then verify directly the validity (10.82).

Remark 10.7. The case where $d > 1$ is more complicated. In fact neither a transformation formula analogous to (10.77) exists in general, nor a formula analogous to (10.82) relating the Green function of the Schrödinger equation with a time-dependent resp. time-independent harmonic oscillator potential (see for instance [398, 399] for some partial results in this direction).

The above results can be generalized to more general time dependent potentials of the following form

$$V(t, x) = V_0(t, x) + V_1(t, x), \tag{10.83}$$

where V_0 is of the type (10.69) or (10.74) and $V_1 : [0, t] \times \mathbb{R}^d \to \mathbb{R}$ satisfies the following assumptions:

1. For each $\tau \in [0, t]$, the application $V_1(\tau, \cdot) : \mathbb{R}^d \to \mathbb{R}$ belongs to $\mathcal{F}(\mathbb{R}^d)$, i.e. $V_1(\tau, x) = \int_{\mathbb{R}^d} e^{ikx} \sigma_\tau(dk)$, $\sigma_\tau \in \mathcal{M}(\mathbb{R}^d)$
2. The application $\tau \in [0, t] \mapsto \sigma_\tau \in \mathcal{M}(\mathbb{R}^d)$ is continuous in the norm $\| \cdot \|$ of the Banach space $\mathcal{M}(\mathbb{R}^d)$

Under the assumptions above it is possible to prove that the application

$$\gamma \in \mathcal{H}_t \mapsto e^{-\frac{i}{\hbar} \int_r^s V(u, \gamma(u) + x) du}, r, s \in [0, t]$$

belongs to $\mathcal{F}(\mathcal{H}_t)$. Moreover by assuming that the initial datum φ belongs to $L_2(\mathbb{R}^d) \cap \mathcal{F}(\mathbb{R}^d)$, the Feynman path integral representation for the solution of the Schrödinger equation with time dependent potential (10.83)

$$\int_{\gamma(t)=0} e^{\frac{i}{2\hbar} \int_0^t |\dot\gamma(\tau)|^2 d\tau - \frac{i}{2\hbar} \int_0^t (\gamma(\tau)+x)\Omega^2(\tau)(\gamma(\tau)+x)d\tau}$$

$$e^{-\frac{i}{\hbar} \int_0^t f(\tau)\cdot(\gamma(\tau)+x)d\tau} e^{-\frac{i}{\hbar} \int_0^t V(\tau,\gamma(\tau)+x)d\tau} \varphi(\gamma(0) + x)d\gamma$$

(with $\Omega^2(\tau) = \Omega^2$ independent of τ if $f \neq 0$) can be rigorously mathematically realized as an infinite dimensional oscillatory integral. More precisely the following holds.

Theorem 10.14. *Under the assumptions above, the following infinite dimensional oscillatory integral on the Cameron-Martin space $\mathcal{H}_t$*

$$\widetilde{\int_{\mathcal{H}_t}^{\circ}} e^{\frac{i}{2\hbar} \int_0^t |\dot\gamma(\tau)|^2 d\tau - \frac{i}{2\hbar} \int_0^t (\gamma(\tau)+x)\Omega^2(\tau)(\gamma(\tau)+x)d\tau}$$

$$e^{-\frac{i}{\hbar} \int_0^t f(\tau)\cdot(\gamma(\tau)+x)d\tau} e^{-\frac{i}{\hbar} \int_0^t V(\tau,\gamma(\tau)+x)d\tau} \varphi(\gamma(0) + x)d\gamma \tag{10.84}$$

is a representation of the solution of Schrödinger equation with time dependent potential V given by (10.83) and initial datum φ.

For a proof see [107].

10.5.3 Phase Space Feynman Path Integrals

Let us recall that Feynman's original aim was to give a Lagrangian formulation of quantum mechanics. On the other hand an Hamiltonian formulation could be preferable from many points of view. For instance the discussion of the semiclassical limit of quantum mechanics is more natural in an Hamiltonian setting (see, e.g., [69, 196, 216, 364] for a discussion of this behavior): in fact the "phase space" rather than the "configuration space" is the natural framework of classical mechanics.

In the physical literature one can often find a "phase space Feynman path integral" representation for the solution of the Schrödinger equation, that is an heuristic formula of the following type:

$$
\text{"}\psi(t,x) = \text{const} \int_{q(t)=x} e^{\frac{i}{\hbar}S_t(q,p)} \varphi(q(0)) dq dp\text{"}.
\tag{10.85}
$$

The integral is meant on the space of paths $(q(\tau), p(\tau))$, $\tau \in [0,t]$ in the phase space of the system $(q(\tau))_{\tau \in [0,t]}$ is the path in configuration space and $p(\tau)_{\tau \in [0,t]}$ is the path in momentum space) and S_t is the action functional in the Hamiltonian formulation:

$$
S_t(q,p) = \int_0^t (\dot{q}(\tau)p(\tau) - H(q(\tau), p(\tau))) d\tau,
$$

(H being the classical Hamiltonian of the system). Different approaches have been proposed for giving a well defined mathematical meaning to the heuristic formula (10.85), for instance via analytic continuation of probabilistic integrals and by considering coherent states [196], or as an "infinite dimensional distribution" (see [169, 170, 171, 172] and references therein). In particular in [80] the phase space Feynman path integral (10.85) has been realized as a well defined infinite dimensional oscillatory integral on a suitable Hilbert space of paths. This approach is particularly advantageous if one is interested in the study of the semiclassical limit by means of a rigorous application of the stationary phase method (adapting the method in [87, 69] as recalled in Sect. 10.3).

Let us introduce the Hilbert space $\mathcal{H}_t \times L_t$, namely the space of paths in the d−dimensional phase space $(q(\tau), p(\tau))_{\tau \in [0,t]}$, such that the path $(q(\tau))_{\tau \in [0,t]}$ belongs to the space of the absolutely continuous functions q from $[0, t]$ to $\mathbb{R}^d$ such that $q(t) = 0$ and $\dot{q} \in L_2([0, t], \mathbb{R}^d)$, while the path in the momentum space $(p(\tau))_{\tau \in [0,t]}$ belongs to $L_t = L_2([0, t], \mathbb{R}^d)$, endowed with the inner product

$$
(q_1, p_1; q_2, p_2) = \int_0^t \dot{q}_1(\tau)\dot{q}_2(\tau) d\tau + \int_0^t p_1(\tau)p_2(\tau) d\tau.
$$

Let us consider the following bilinear form:

$$
[q_1, p_1; q_2, p_2] =
$$
$$
\int_0^t \dot{q}_1(\tau)p_2(\tau) d\tau + \int_0^t p_1(\tau)\dot{q}_1(\tau) d\tau - \int_0^t p_1(\tau)p_2(\tau) d\tau = (q_1, p_1; B(q_2, p_2)),
\tag{10.86}
$$

where B is the following operator in $\mathcal{H}_t \times L_t$,:

$$B(q,p)(\tau) = (\int_t^\tau p(u)du, \dot{q}(\tau) - p(\tau)). \qquad (10.87)$$

B is densely defined, e.g. on $C^1([0,t];\mathbb{R}^d) \times C^1([0,t];\mathbb{R}^d)$, it is invertible with inverse given by

$$B^{-1}(q,p)(\tau) = (\int_t^\tau p(u)du + q(\tau), \dot{q}(\tau)) \qquad (10.88)$$

(on the range of B). The quadratic form on $\mathcal{H}_t \times L_t$ given by $(q,p) \mapsto ((q,p), B(q,p)) = 2\int_0^t \dot{q}(\tau)p(\tau)d\tau - \int_0^t p(\tau)^2 d\tau$ can be recognized as the classical Hamiltonian action of the system $S(q,p)$ (multiplied by 2) along the path (q,p) for the free particle, i.e. $H(q,p) = p^2/2$. The phase space Feynman path integral representation, or in other words the integration with respect to the phase space variables, becomes particularly interesting when the potential depends explicitly both on position and on velocity, i.e. $H(q,p) = p^2/2 + V(q,p)$. Under suitable assumptions on the function V and the initial datum φ it is possible to realize the heuristic expression (10.85) as an infinite dimensional oscillatory integral (in the sense of Definition 10.4):

Theorem 10.15. *Let us consider the following Hamiltonian[1]*

$$H(Q;P) = \frac{P^2}{2} + V_1(Q) + V_2(P)$$

in $L^2(\mathbb{R}^d)$ and the corresponding Schrödinger equation

$$\begin{cases} \dot{\psi} = -\frac{i}{\hbar}H\psi \\ \psi(0,x) = \varphi(x), \quad x \in \mathbb{R}^d \end{cases} \qquad (10.89)$$

Let us assume that $V_1, \varphi \in \mathcal{F}(\mathbb{R}^d)$ and $\int_0^t V_2(p(\tau))d\tau \in \mathcal{F}(L_t)$. Then the functional

$$f(q,p) = \varphi(x + q(0))e^{-\frac{i}{\hbar}\int_0^t V(q(s)+x,p(s))ds}$$

belongs to $\mathcal{F}(\mathcal{H}_t \times L_t)$ and the well defined normalized Fresnel integral with respect to the operator B in (10.87)

$$\widetilde{\int_{\mathcal{H}_t \times L_t}^B} e^{\frac{i}{2\hbar}(q,p;B(q,p))}e^{-\frac{i}{\hbar}\int_0^t(V_1(q(\tau)+x)+V_2(p(\tau)))d\tau}\varphi(q(0)+x)dqdp$$

is a representation of the solution of the Cauchy problem (10.89).

For a proof of these results see [80]. For a different application of phase space Feynman path integral see [452] and for a different approach [287].

---

[1] The general case presents problems due to the non commutativity of the quantized expression of Q and P, for a different approach with more general Hamiltonians see [420].

10.5.4 The Stochastic Schrödinger Equation

A recent and particularly interesting application of Feynman path integrals can be found in the quantum theory of measurement. Let us recall that the continuous time evolution of a quantum system described by the traditional Schrödinger equation (1.3) is valid if the quantum system is "undisturbed", but if it is submitted to the measurement of one of its observables the influence on the measuring apparatus on it cannot be neglected. In fact the state of the system after the measurement is the result of a random and discontinuous change, the so-called "collapse of the wave function", which cannot be described by the ordinary Schrödinger equation. There are several efforts to include the process of measurement into the traditional quantum theory and to deduce from its laws, instead of postulating, the collapse of the wave function. In particular the aim of the *quantum theory of measurement* is a description of the process of measurement taking into account the properties of the measuring apparatus, which is handled as a quantum system, and its interaction with the system submitted to the measurement [201].

An alternative Feynman path integral approach was proposed by Mensky [373]. In fact he proposed an heuristic formula for the selective dynamics of a particle whose position is continuously observed. According to Mensky the state of the particle at time t if the observed trajectory is the path $\omega(s)_{s\in[0,t]}$ is given by the "restricted path integral"

$$\psi(t,x,\omega) = \text{``} \int_{\{\gamma(t)=x\}} e^{\frac{i}{\hbar}S_t(\gamma)} e^{-\lambda \int_0^t (\gamma(s)-\omega(s))^2 ds} \varphi(\gamma(0)) D\gamma \text{ ''}, \quad (10.90)$$

where $\varphi \in L_2(\mathbb{R}^d)$ is the initial state of the system, S_t is the action functional and $\lambda > 0$ a real positive parameter. One can see that, as an effect of the correction term $e^{-\lambda \int_0^t (\gamma(s)-\omega(s))^2 ds}$ due to the measurement, the paths γ giving the main contribution to the integral (10.90) are, heuristically, those closer to the observed trajectory ω.

Another alternative phenomenological description for the process of "unsharp" continuous quantum measurement can be given by means of a class of stochastic Schrödinger equations, see for instance [144, 131, 132, 212, 373, 248, 134, 135, 136, 322, 57, 159]. A particular example is Belavkin's equation, a stochastic Schrödinger equation describing the selective dynamics of a $d-$dimensional particle submitted to the measurement of one of its (possible $M-$dimensional vector) observables, described by the self-adjoint operator R on $L^2(\mathbb{R}^d)$

$$\begin{cases} d\psi(t,x) = -\frac{i}{\hbar}H\psi(t,x)dt - \frac{\lambda}{2}R^2\psi(t,x)dt + \sqrt{\lambda}R\psi(t,x)dW(t) \\ \psi(0,x) = \varphi(x) \qquad (t,x) \in [0,T] \times \mathbb{R}^d \end{cases} \quad (10.91)$$

where H is the quantum mechanical Hamiltonian, W is an $M-$dimensional Brownian motion on a probability space $(\Omega, \mathcal{F}, \mathbb{P})$, $dW(t)$ is the Ito differential and $\lambda > 0$ is a coupling constant, which is proportional to the accuracy of the measurement. In the particular case of the description of the continuous measurement of position one has $R = x$ (the multiplication operator in $L_2(\mathbb{R}^d)$, so that equation (10.91) assumes the following form:

$$
\begin{cases}
\mathrm{d}\psi(t,x) = -\frac{\mathrm{i}}{\hbar}H\psi(t,x)\mathrm{d}t - \frac{\lambda}{2}x^2\psi(t,x)\mathrm{d}t + \sqrt{\lambda}x\psi(t,x)\mathrm{d}W(t) \\
\psi(0,x) = \varphi(x) \qquad (t,x) \in [0,T] \times \mathbb{R}^d,
\end{cases}
\tag{10.92}
$$

while in the case of momentum measurement, $(R = -\mathrm{i}\hbar\nabla)$, one has:

$$
\begin{cases}
\mathrm{d}\psi(t,x) = -\frac{\mathrm{i}}{\hbar}H\psi(t,x)\mathrm{d}t + \frac{\lambda\hbar^2}{2}\triangle\psi(t,x)\mathrm{d}t - \mathrm{i}\sqrt{\lambda}\hbar\nabla\psi(t,x)\mathrm{d}W(t) \\
\psi(0,x) = \psi_0(x) \qquad (t,x) \in [0,T] \times \mathbb{R}^d.
\end{cases}
\tag{10.93}
$$

·Belavkin derives (10.91) by modeling the measuring apparatus (but it would be better to say "the informational environment") by means of a one-dimensional bosonic field and by assuming a particular form for the interaction Hamiltonian between the field and the system on which the measurement is performed. The solution ψ of the Belavkin equation is a stochastic process, whose expectation values have an interesting physical meaning. Let $w(s), s \in [0,t]$ be a continuous path (from $[0,t]$ into $\mathbb{R}^M$), I a Borel set in the Banach space $C([0,t],\mathbb{R}^M)$ endowed with the sup norm, and let P be the Wiener measure on $C([0,t],\mathbb{R}^M)$. The probability that the observed trajectory up to time t, i.e. the values of the observable $R(s)_{s\in[0,t]}$, belongs to the set of paths I is given by the following Wiener integral:

$$
\mathbb{P}(R(s) = w(s)_{s\in[0,t]} \in I) = \int_I |\psi(t,w)|^2 \mathrm{d}P(w).
$$

Moreover if we measure at time t another observable of the system, denoted with Z, then its expected value, conditioned to the information that the observed trajectory of R up to time t belongs to the Borel set I, is given by:

$$
\mathbb{E}(Z(t)|R(s) = w(s)_{s\in[0,t]} \in I) = \int_I \frac{\langle \psi(t,w), Z\psi(t,w)\rangle}{|\psi(t,w)|^2} \mathrm{d}P(w).
$$

(where $\psi(t,w) \neq 0$ is assumed).

Infinite dimensional oscillatory (Fresnel) integrals provide a Feynman path integral representation of the solution of the stochastic Schrödinger equations (10.92) and (10.93) and, as a consequence, also a rigorous mathematical realization of the heuristic formula (10.90) [96, 97, 450, 451, 81, 82]. The present result is a generalization of Theorem 10.1 to complex valued phase functions.

Theorem 10.16 ([81]). *Let $\mathcal{H}$ be a real separable Hilbert space, let $y \in \mathcal{H}$ be a vector in $\mathcal{H}$ and let L_1 and L_2 be two self-adjoint, trace class commuting operators on $\mathcal{H}$ such that $I + L_1$ is invertible and L_2 is non negative. Let moreover $f : \mathcal{H} \to \mathbb{C}$ be the Fourier transform of a complex bounded variation measure μ_f on $\mathcal{H}$:*

$$f(x) = \hat{\mu}_f(x), \qquad f(x) = \int_{\mathcal{H}} e^{i(x,k)} d\mu_f(k), \qquad x \in \mathcal{H}.$$

Then the function $g : \mathcal{H} \to \mathbb{C}$ given by

$$g(x) = e^{\frac{i}{2\hbar}(x,Lx)} e^{(y,x)} f(x)$$

(L being the operator on the complexification $\mathcal{H}^{\mathbb{C}}$ of the real Hilbert space $\mathcal{H}$ given by $L = L_1 + iL_2$) is integrable (in the sense of Definition 10.3) and its infinite dimensional oscillatory integral

$$\widetilde{\int_{\mathcal{H}}} e^{\frac{i}{2\hbar}(x,(I+L)x)} e^{(y,x)} f(x) dx$$

can be explicitly computed by means of the following Parseval type equality:

$$\widetilde{\int_{\mathcal{H}}} e^{\frac{i}{2\hbar}(x,(I+L)x)} e^{(y,x)} f(x) dx = \det(I+L)^{-1/2} \int_{\mathcal{H}} e^{\frac{-i\hbar}{2}(k-iy,(I+L)^{-1}(k-iy))} d\mu_f(k)$$

$$(10.94)$$

The following theorem gives an application of Theorem 10.16 to the solution of the stochastic Schrödinger equation:

Theorem 10.17 ([81]). *Let V and φ be Fourier transform of finite complex measures on $\mathbb{R}^d$. Then there exists a (strong) solution of the stochastic Schrödinger equation (10.92) and it can be represented by the following infinite dimensional oscillatory integral with complex phase on the Hilbert space $(\mathcal{H}_t, (,))$ of absolutely continuous function $\gamma : [0,t] \to \mathbb{R}^d$ such that $\int_0^t |\dot{\gamma}(s)|^2 ds < \infty$, with inner product $(\gamma_1, \gamma_2) = \int_0^t \dot{\gamma}_1(s)\dot{\gamma}_2(s) ds$:*

$$\psi(t,x) = \widetilde{\int} e^{\frac{i}{\hbar} S_t(\gamma) - \lambda \int_0^t (\gamma(s)+x)^2 ds} e^{\int_0^t \sqrt{\lambda}(\gamma(s)+x) dW(s) ds} \varphi(\gamma(0) + x) d\gamma$$

$$= e^{-\lambda|x|^2 t + \sqrt{\lambda} x \cdot \omega(t)} \widetilde{\int_{\mathcal{H}_t}} e^{\frac{i}{2\hbar}(\gamma,(I+L)\gamma)} e^{(l,\gamma)} e^{-2\lambda\hbar \int_0^t x \cdot \gamma(s) ds}$$

$$\cdot e^{-i \int_0^t V(x+\gamma(s)) ds} \varphi(\gamma(0) + x) d\gamma$$

where $l \in \mathcal{H}_t$, $l(s) = \sqrt{\lambda} \int_s^t \omega(\tau) d\tau$ and

$$L : \mathcal{H}_t^{\mathbb{C}} \to \mathcal{H}_t^{\mathbb{C}}, \qquad (\gamma_1, L\gamma_2) = -2i\lambda\hbar \int_0^t \gamma_1(s)\gamma_2(s) ds$$

The existence and uniqueness of a strong solution of (10.92) is proved in [248]. For a detailed proof of the Feynman path integral representation (Theorem 10.17) see [81, 371]).

A corresponding result can be obtained for the Belavkin equation (10.93) describing a continuous momentum measurement [82, 450, 451].

10.5.5 The Chern–Simons Functional Integral

A particularly interesting application of heuristic Feynman path integrals can be found in a paper by Witten [465], who conjectured that there should be a connection between quantum field theories based on the so-called Chern–Simons action and the Jones polynomial, a link invariant (see also [127, 160, 262, 362, 237]). (Witten's work was preceded, in the abelian case, by work of Schwarz [407, 408]). Witten's heuristic calculations are based on a heuristic Feynman path integral formulation of Chern–Simons theory, where the integration is performed on a space of geometric objects, i.e. on a space of connections. Chern–Simons theory is a quantum gauge field theory in 3 dimensions (Euclidean 3-dimensional space-time). More precisely let M be a smooth 3-dimensional oriented manifold without boundary, let G be a compact Lie group (the "gauge group"), and let g be its Lie algebra with a fixed Ad-invariant inner product $\langle \cdot, \cdot \rangle$. Let A be a g–valued connection 1-form and let

$$S_{CS}(A) \equiv \frac{k}{4\pi} \int_M \left(\langle A \wedge dA \rangle + \frac{1}{3} \langle A \wedge [A \wedge A] \rangle \right),$$

be the Chern–Simons action, where k is a non-zero real constant, $[A, A]$ is the 2-form whose value on a pair of vectors (X, Y) is $2[A(X), A(Y)]$ and $\langle A \wedge B \rangle$, for a g–valued 1-form A and a g–valued 2-form B is the 3-form whose value on (X, Y, Z) is given by the skew-symmetrized form of $\langle A(X), B(Y, Z) \rangle$.

According to Witten–Schwarz's conjecture, the heuristic Feynman path measure of the form

$$\mathrm{d}\mu_F(A) \equiv \frac{1}{Z} \mathrm{e}^{\mathrm{i}S_{CS}(A)} DA, \tag{10.95}$$

(where DA is a heuristic flat measure on the space $\mathcal{A}_M$ of smooth g–valued connection 1-forms on M with compact support and Z is a normalization constant) should allow for the computation of topological invariants of the manifold M. Indeed let L be a link in M, i.e. a n-tuple $(l_1, ..., l_n)$, $n \in \mathbb{N}$, of loops in M whose arcs are pairwise disjoint and let $WLF(L)$ be the Wilson loop function associated to L, that is the map:

$$WLF(L) : \mathcal{A}_M \ni A \mapsto \prod_{i=1}^{n} \mathrm{Tr}(Hol(A, l_i)) \in \mathbb{C},$$

where $Hol(A, l)$ denotes the holonomy of A around l. The heuristic integral

$$WLO(L) := \int WLF(L) \mathrm{d}\mu_F \tag{10.96}$$

is called Wilson loop observable associated to the link L. According to Witten the function mapping every sufficiently regular link L to the associated Wilson loop observable $WLO(L)$ should be a link invariant.

A rigorous formulation of expression (10.96) and a justification of Witten–Schwarz's conjecture were obtained in the case G is abelian by Albeverio and Schäfer [405, 115] in terms of Fresnel integrals, and in terms of white noise analysis by Leukert and Schäfer [353]. This result was generalized to the non abelian case for $M = \mathbb{R}^3$ in [116] by means of white noise analysis (see also [83] for a detailed exposition of this topic). Recently rigorous results on an asymptotic expansion of (a regularization of) (10.96) in powers of k have been obtained in [112].

We present here the result of [116], where a suitable choice of gauge leads to a quadratic expression for the transformed S_{CS}. In fact every connection on $\mathbb{R}^3$ can be gauge transformed into the form

$$A = a_0 dx_0 + a_1 dx_1,$$

where a_0 and a_1 are functions over $\mathbb{R}^3$ taking values in the Lie algebra g, and

$$a_{1|(\mathbb{R}^2 \times \{0\})} = 0, \qquad a_{0|(\mathbb{R} \times \{(0,0)\})} = 0. \qquad (10.97)$$

In this gauge the Chern–Simons action S_{CS} loses the cubic term and, after integration by parts, becomes

$$S_{CS}(a_0, a_1) = \frac{k}{2\pi} \langle a_0, -\partial_2 a_1 \rangle_{L^2(\mathbb{R}^3, g)}$$

and the Chern–Simons integrals take the form

$$\int_{\mathcal{A}'} f(a_0, a_1) e^{iS_{CS}(a_0, a_1)} da_0 da_1, \qquad (10.98)$$

where the integral is meant on the space $\mathcal{A}'$ of connections (a_0, a_1) satisfying (10.97). Instead of (a_0, a_1) we shall use (a_0, f_1), with $f_1 = \partial_2 a_1$. With this convention the Chern–Simons action assumes the simple form $S_{CS}(a_0, f_1) = \frac{k}{2\pi} \langle a_0, f_1 \rangle_{L^2(\mathbb{R}^3, g)}$. Expression (10.98) can be rigorously mathematically realized by means of white noise analysis (see Sect. 10.4.2), by considering the Hilbert space $E := L^2_{real}(\mathbb{R}^3; g \oplus g) \simeq (L^2_{real}(\mathbb{R}^3) \otimes g) \oplus (L^2_{real}(\mathbb{R}^3) \otimes g)$. $L^2_{real}(\mathbb{R}^3)$ denotes the space of real-valued functions which are square integrable with respect to the Lebesgue measure on $\mathbb{R}^3$. The operator K in (10.55) is taken to be the identity on $g \oplus g$ and is given by $K_1^{\otimes 3}$ on $L^2_{real}(\mathbb{R})^{\otimes 3} \simeq L^2_{real}(\mathbb{R}^3)$, where K_1 is the operator $\frac{1}{4}(-\frac{d^2}{dx^2} + \frac{x^2}{4})^{-1}$ on $L^2_{real}(\mathbb{R})$. $\mathcal{E}_p$ is the space

$$\{(K^p a_0, K^p f_1) : a_0, f_1 \in L^2_{real}(\mathbb{R}^3) \otimes g\}$$

and the $\langle \cdot, \cdot \rangle_p$ inner product is specified by the norm

$$\|(a_0, f_1)\|_p^2 = \|K^{-p} a_0\|_{L^2(\mathbb{R}^3, g)}^2 + \|K^{-p} f_1\|_{L^2(\mathbb{R}^3, g)}^2.$$

The Chern–Simons integrator is defined as the distribution Φ_{CS} on $\mathcal{E}^* = S'_{g \oplus g}(\mathbb{R}^3)$ whose S-transform (an infinite dimensional analogue of the finite dimensional Laplace transform, relative to Gaussian measure rather than Lebesgue measure, see Sect. 10.4.2) is

$$S\Phi_{CS}(j_0, j_1) = e^{\frac{2\pi}{k} i(j_0, j_1)_{L^2(\mathbb{R}^3, g)} - ((j_0, j_1), (j_0, j_1))_0/2}.$$

The construction of observables is rather involved. Indeed for general elements A in the domain of Φ_{CS} the expression $Hol(A, l)$ as well as the function $WLF(L)$ make no sense. The problem can be solved considering a "smeared" version $WLF(L, \epsilon)$ of $WLF(L)$, defined by replacing all expressions $l(t)$, $t \in S^1$, with functions $l^\epsilon(t)$ concentrated in a ϵ-neighborhood of l, given by $l^\epsilon(t) := \psi^\epsilon((\cdot) - l(t))$, where $\psi^\epsilon := \frac{1}{\epsilon^3} \psi(\frac{(\cdot)}{\epsilon})$ and ψ is a element of $C^\infty(\mathbb{R}^3)$ such that $\psi \geq 0$, $\mathrm{supp}(\psi) \in B_1(0)$ and $\int_{\mathbb{R}^3} \psi(x) dx = 1$. The function $WLF(L, \epsilon)$ so constructed is measurable on $\mathcal{E}^*$ and belongs to $[\mathcal{E}]$, so that $\Phi_{CS}(WLF(L, \epsilon))$ is well defined. Unfortunately for the definition of $\Phi_{CS}(WLF(L))$ it is not sufficient to take $\lim_{\epsilon \to 0} \Phi_{CS}(WLF(L, \epsilon))$, as this does not exist for all L belonging to a sufficiently large set of links in $\mathbb{R}^3$ and the mapping

$$L \mapsto \lim_{\epsilon \to 0} \Phi_{CS}(WLF(L, \epsilon)) \in \mathbb{C}$$

is not, in general, a link invariant. One has to introduce an additional regularization: the so called "framing procedure". This is done by choosing a suitable family $(\varphi_s)_{s>0}$ of diffeomorphisms of $\mathbb{R}^3$ such that $(\varphi_s \circ l_k)_{s>0}$ approximates the loop l_k for every $k \leq n$ and compute $WLO(L, \epsilon; \varphi_s) := \Phi_{CS}(WLF(L, \epsilon; \varphi_s))$. It is possible to prove (see [266, 267, 268] for details) that if the framing $(\varphi_s)_{s>0}$ and the link $L = (l_1, ..., l_n)$ satisfy suitable assumptions, the limit

$$WLO(L, \varphi_s) := \lim_{s \to 0} \lim_{\epsilon \to 0} WLO(L, \epsilon; \varphi_s)$$

exists and represents a topological invariant (see [83, 268, 269] for a detailed exposition). However the values of the WLOs obtained by means of this procedure do not coincide with the values presented in various publications in the physical literature, even if they are quite similar. In [269] Hahn conjectured that the main reason for this non coincidence is the non compactness of the manifold $M = \mathbb{R}^3$. In a recent paper [270] the same author studies the case of a compact product manifolds M of the form $M = \Sigma \times S^1$, where Σ is an oriented surface, by applying special gauge fixing procedures, called "quasi-axial gauge" resp "torus gauge". By using suitable regularization procedures as "loop smearing" and "framing", the values of the WLOs can be rigorously computed, both for abelian and non-abelian groups G (see [270, 271, 272] for a detailed exposition).

Notes

Section 10.1 An other alternative approach to the rigorous mathematical definition of Feynman path integrals makes use of nonstandard analysis [78]. The starting point is the finite dimensional approximation of the Feynman path integral by means of piecewise linear paths, as explained in Chap. 1, (1.10):

$$e^{-\frac{i}{\hbar}tH}\varphi(x) = s - \lim_{k\to\infty}\left(2\pi i\frac{\hbar}{m}\frac{t}{k}\right)^{-\frac{kn}{2}}\int_{\mathbb{R}^{nk}} e^{-\frac{i}{\hbar}S_t(x_k,\ldots,x_0)}\varphi(x_0)dx_0\ldots dx_{k-1}$$

The leading idea is the extension of the expression

$$\left(2\pi i\frac{\hbar}{m}\frac{t}{k}\right)^{-\frac{kn}{2}}\int_{\mathbb{R}^{nk}} e^{-\frac{i}{\hbar}S_t(x_k,\ldots,x_0)}\varphi(x_0)dx_0\ldots dx_{k-1}$$

from $k \in \mathbb{N}$ to a nonstandard hyperfinite infinite $k \in {}^*\mathbb{N}$. The result is just an *internal* quantity. For a suitable class of potentials its standard part can be shown to exist and to solve the Schrödinger equation.

In an analogous way the semiclassical approximation of the solution of the Schrödinger equation can be obtained by taking the parameter $\hbar$ as an infinitesimal quantity.

Even if this approach provides a very suggestive realization of the Feynman path integrals, it has not been systematically developed yet. For a more detailed description of non standard methods in mathematical physics see [78] and references therein. For a newer development see [380, 379, 381, 356, 357, 358].

Section 10.2 The problem of defining rigorously Feynman path integrals for Schrödinger operators which are polynomial (or have polynomial growth) of order greater than 3 has been open since the very introduction of Feynman path integrals. A "time slicing" or "sequential approach" is possible in the form of a Lie–Trotter–Kato formula (see Sect. 10.4.3 and, e.g., [395, 383]), however a rigorous study of the semiclassical limit seems to be absent. The approach by Fujiwara and Kumano-Go of piecewise classical paths does not extend beyond potentials with at most quadratic growth at infinity [239, 240, 241, 242, 243, 246, 244, 335]. The approach by Laplace-transform does allow special potentials of exponential growth, but also does not cover polynomial potentials [73, 92, 336]. Rather indirect approaches to polynomial potentials, involving approximations and analytic continuations, which seems to make difficult any semiclassical expansion are in [411, 421, 137, 138, 139, 140]. For semiclassical expansions for eigenvalues and time independent eigenfunctions based on the corresponding equations with polynomial potentials see e.g. [413, 205]. The approach developed in this section is based on [69, 223, 105].

Section 10.3 Semiclassical expansions constitute a basic chapter in quantum theory, related as they are with the correspondence principle [338, 155, 363], approximate calculations [355, 474], quantization procedures [382], see also, e.g., [77, 181, 182, 215, 224, 225, 245, 375, 415, 448, 452, 454, 455, 456, 457]. They are also used as heuristic tools in low dimensional topology and differential geometry [127, 268, 269, 112, 124]. They are also related to ray optics and certain high frequency approximations in electromagnetism [124, 220, 153, 154], as well as high Reynolds numbers expansions in hydrodynamics [62, 334, 79, 453]. There exists results in quantum mechanics and in the above related areas based on a rigorous version of the WKB method and the Fourier integral operators [220, 141, 162, 273, 158, 189, 400]. Rigorous results based on Feynman path integrals were first obtained by Albeverio and Høegh-Krohn during the writing of the first edition of the present book, but were not included in it but rather published separately [87]. Further results on this line are connected with the trace formula for Schrödinger operators [64, 65, 66, 67], which unfortunately seem to have escaped attention, despite the strong resonance obtained by the related heuristic Gutzwiller trace formula, also developed at about the same time [263, 406].

Asymptotics of eigenvalues and time independent eigenfunctions can be obtained for the corresponding heat equation, see [413, 204, 205, 215, 216, 221]. The trace formula has interesting relations with certain problems of number theory, which deserve further attention, see [91, 66, 67, 110]. See also the comment at the end of section 10.4.1, concerning asymptotic expansions of infinite dimensional probabilistic integrals.

Section 10.4 In this section we only described to a certain extent a few alternative approaches. There are other ones. One has been presented in [420], we already mentioned it in the notes of Sect. 10.2. A combinatorial resp. discrete approach has been presented in [191] resp [156]. An approach by Daubechies and Klauder [196, 197, 198, 199, 200] concerns phase space integrals in quantum theory. The semiclassical limit has not been discussed in it. There are approaches based on "exact formulae" for special potentials. They are described extensively in several books [255, 256, 319, 406] (the reader should be aware that in these references not all formulae are proven rigorously, some are only suggested on the basis of heuristic arguments resp. on analogies with formulae in a corresponding probabilistic setting). Heuristic formulae inspired by Feynman path integrals have also been widely used in quantum theory in general [319, 406, 464] and especially in quantum field theory [164, 174, 296, 317, 389, 394, 462, 463, 475]. In particular those connected with instanton expansions and of saddle point methods are of this type (and heuristically related to the heat equation and associated probabilistic integrals). In addition there are formulae related to critical point expansions. In the last two decades such formulae have also given a lot of inspiration to other areas of mathematics, through work by Atiyah and Witten, as well as work by e.g. Duistermaat, Donaldson, Kontsevich, Polyakov.

There has often been, on the other side, an unfortunate tendency of completely ignoring rigorization of path integrals, for the sake of "quick insight" or progress. Even though this might be understandable for a time, it would certainly be negative to mathematics in the long run, if it would develop into a systematic attitude. It is in any case obvious that this is not the attitude of the present book, which insists on the need of rigorous approaches. Let us finally mention that there are mathematical approaches to supersymmetric Feynman path integrals, see, e.g., [349, 350, 352, 292, 419, 402].

Section 10.5 Here we only describe a few applications of Feynman path integrals. There are other ones, as e.g. the application to the solution of Dirac equation [283, 288, 380, 381, 410, 468, 469, 470, 471, 157]. A non exhaustive list of further applications and topics includes Feynman path integrals with singular potentials [133, 235, 249, 252], relations to fractals [339, 340, 341, 342, 343, 431] path integrals on manifolds [293, 294, 349, 350, 351, 352, 434, 441] resp. on other non commutative structures [417], wave equations and related path integrals [181, 182, 344, 345, 346, 347], hyperbolic systems [437, 438], quantum statistical systems [430].

The approach to rigorous Chern–Simons integrals which we present here has been developed in [115, 116, 117]. Sketches for an axiomatic approach have been given in [310], see also [311, 312, 313].

References of the First Edition

1. R.P. Feynman, Space-time approach to non-relativistic quantum mechanics, Rev. Mod. Phys. 20, 367–387 (1948).
(Based on Feynman's Princeton Thesis, 1942, recently published as *Feynman's Thesis - a new approach to quantum theory*, edited by Laurie M. Brown (Northwestern University, USA, 2005)).
The first applications to quantum field theory are in
R.P. Feynman, The theory of positrons, Phys. Rev. 76, 749–759 (1949); Space-time approach to quantum electro-dynamics, Phys. Rev. 76, 769–789 (1949); Mathematical formulation of the quantum theory of electromagnetic interaction, Phys. Rev. 80, 440–457 (1950).
(All quoted papers are also reprinted in: J. Schwinger, Selected Papers on Quantum Electrodynamics, Dover, New York (1958).
See also: R.P. Feynman, A.R. Hibbs, Quantum mechanics and path integrals. MacGraw Hill, New York (1965).
R.P. Feynman, The concept of probability in quantum mechanics, Proc. Second Berkeley Symp.Math. Stat. Prob., Univ. Calif. Press (1951), pp. 533–541.
2. P.A.M. Dirac, The Lagrangian in quantum mechanics, Phys. Zeitschr. d. Sowjetunion, 3, No 1, 64–72 (1933).
P.A.M. Dirac, The Principles of Quantum Mechanics, Clarendon Press, Oxford, IV. Ed. (1958), Ch. V, §32, p. 125.
P.A.M. Dirac, On the analogy between classical and quantum mechanics, Rev. Mod. Phys. 17, 195–199 (1945)
3. R.P. Feynman, The development of the space-time view of quantum electrodynamics, in Nobel lectures in Physics, 1965, Elsevier Publ. (1972).
4. 1) F. Dyson, Missed opportunities, Bull. Amer. Math. Soc. 78, 635–652 (1972).
    2) Also e.g. L .D. Faddeev, The Feynman integral for singular Lagrangians, Teor. i Matem. Fizika, 1, 3–18 (1969) (transl. Theor. Mathem. Phys. 1, 1–13 (1969)).
5. I.M. Gelfand, A.M. Yaglom, Integration in functional spaces, J. Math. Phys. 1, 48–69 (1960) (transl. from Usp. Mat. Nauk. 11, Pt. 1; 77–114 (1956)).
6. Exclusively concerned with reviewing work on specifically mathematical problems:

1) E.J. McShane, Integral devised for special purposes, Bull. Am. Math. Soc. 69, 597–627 (1963).

2) L. Streit, An introduction to theories of integration over function spaces, Acta Phys. Austr. Suppl. 2, 2–20 (1966).

3) Yu L. Daletskii, Integration in function spaces, in Progress in Mathematics, Vol. 4, Ed. R.V. Gamkrelidze (transl.), Plenum, New York (1969), 87–132.

The following reviews are somewhat less concentrated on specific mathematical problems, and include also some discussions of heuristic work:

4) E.W. Montroll, Markoff chains, Wiener integrals and quantum theory, Commun. Pure Appl. Math. 5, 415–453 (1952).

5) I.M. Gelfand, A.M. Yaglom, see [5].

6) S.G. Brush, Functional integrals and statistical physics, Rev. Mod. Phys. 33, 79–92 (1961).

7) C. Morette DeWitt, L'intégrale fonctionnelle de Feynman. Une introduction, Ann. Inst. Henri Poincaré, 11, 153–206 (1969).

8) J. Tarski, Definitions and selected applications of Feynman-type integrals, Presented at the Conference on functional integration at Cumberland Lodge near London, 2–4 April, 1974, Hamburg Universität, Preprint, May 1974.

7. 1) C. Morette, On the definition and approximation of Feynman's path integrals, Phys. Rev. 81, 848–852 (1951).

2) W. Pauli, Ausgewählte Kapitel aus der Feldquantisierung, ausg. U. Hochstrasser, M.F. Schafroth, E.T.H., Zürich, (1951), Appendix.

3) Ph. Choquard, Traitement semi-classique des forces générales dans la représentation de Feynman. Helv. Phys. Acta 28, 89–157 (1955).

Textbooks in which there is some heuristic discussion of the Feynman path integral are e.g.:

4) G. Rosen, Formulation of classical and quantum dynamical theory, New York (1969), Academic Press.

5) A. Katz, Classical mechanics, quantum mechanics, field theory, Academic Press, New York (1965).

6) R. Hermann, Lectures in Mathematical Physics, Benjamin, Reading (1972), Vol. II, Ch. IV.

7) W. Yourgrau, S. Mandelstam, Variational principles in dynamics and quantum theory, Pitman, London (1968).

8) B. Kursunoglu, Modern Quantum Theory, Freeman, San Francisco (1962).

9) D. Lurié, Particles and fields, Interscience, New York (1968).

8. N. Wiener, A. Siegel, B. Rankin, W. Ted Martin, Differential space, quantum systems and prediction, M.I.T. Press (1966).

E. Nelson, Dynamical Theories of Brownian Motion, Princeton Univ. Press. (1967).

J. Yeh, Stochastic processes and the Wiener integral, Dekker, New York (1973).

9. M. Kac, On distributions of certain Wiener functionals, Trans. Amer. Math. Soc. 65, 1–13 (1949).

Also e.g. M. Kac, On some connections between probability theory and differential and integral equations, Proc. Second Berkeley Symp., Univ. California Press, Berkeley (1951) 189–215.

M. Kac, Probability and related topics in physical sciences, Interscience Publ., New York (1959).

10. 1) R.H. Cameron, A family of integrals serving to connect the Wiener and Feynman integrals, J. Math. and Phys. 39, 126–141 (1960).

2) Yu L. Daletskii, Functional integrals connected with operator evolution equations, Russ. Math. Surv. 17, No 5, 1–107 (1962) (transl.)

3) R.H. Cameron, The Ilstow and Feynman integrals, J. D'Anal. Math. 10, 287–361 (1962–63).

4) J. Feldman, On the Schrödinger and heat equations, Trans. Am. Math. Soc. 10, 251–264 (1963).

5) D.G. Babbitt, A summation procedure for certain Feynman integrals, J. Math. Phys. 4, 36–41 (1963).

6) E. Nelson, Feynman integrals and the Schrödinger equation, J. Math. Phys. 5, 332–343 (1964).

7) D.G. Babbitt, The Wiener integral and the Schrödinger equation, Trans. Am. Math. Soc. 116, 66–78 (1965). Correction in Trans. Am. Math. Soc. 121, 549–552 (1966).

8) J.A. Beekman, Gaussian process and generalized Schrödinger equations, J. Math. and Mech. 14, 789–806 (1965).

9) R.H. Cameron, D.A. Storvick, A translation theorem for analytic Feynman integrals, Trans. Am. Math. Soc. 125, 1–6 (1966).

10) D.G. Babbitt, Wiener integral representations for certain semigroups which have infinitesimal generators with matrix coefficients, Bull. Am. Math. Soc. 73, 394–397 (1967).

11) J.A. Beekman, Feynman-Cameron integrals, J. Math. and Phys. 46, 253–266 (1967).

12) R.H. Cameron, Approximation to certain Feynman integrals, Journal d'Analyse Math. 21, 337–371 (1968).

13) R.H. Cameron, D.A. Storvick, An operator valued function space integral and a related integral equation, J. Math. and Mech. 18, 517–552 (1968).

14) R.H. Cameron, D.A. Storvick, A Lindeløf theorem and analytic continuation for functions of several variables, with an application to the Feynman integral, Proc. Symposia Pure Mathem., XI, Am. Math. Soc., Providence (1968), 149–156.

15) J.A. Beckman, Green's functions for generalized Schrödinger equations, Nagoya Math. J. 35, 133–150 (1969). Correction in Nagoya Math. J. 38, 199 (1970).

16) D.L. Skoug, Generalized Ilstow and Feynman integrals, Pac. Journ. of Math. 26, 171–192 (1968).

17) G.W. Johnson, D.L. Skoug, Operator-valued Feynman integrals of certain finite-dimensional functionals, Proc. Am. Math. Soc. 24, 774–780 (1970).

18) G.W. Johnson, D.L. Skoug, Operator-valued Feynman integrals of finite-dimensional functionals. Pac Journ. of Math. 415–425 (1970).

19) G.W. Johnson, D.L. Skoug, An operator valued function space integral: A sequel to Cameron and Storvick's paper, Proc. Am. Math. Loc. 27, 514–518 (1971).

20) J.A. Beekman, Sequential gaussian Markov integrals, Nagoya Math. J. 42, 9–21 (1971).

21) G.W. Johnson, D.L. Skoug, Feynman integrals of non-factorable finite-dimensional functionals, Pac. Journ. of Math. 45, 257–267 (1973).

22) J.A. Beekman, R.A. Kallman, Gaussian Markow expectation and related integral equations, Pac. Journ. Math. 37, 303–317 (1971).

23) G.W. Johnson, D.L. Skoug, A Banach algebra of Feynman integrable functionals with application to an integral equation formally equivalent to Schrödinger's equation, J. Funct. Analysis 12, 129–152 (1973).

24) D.L. Skoug, Partial differential Systems of generalized Wiener and Feynman integrals, Portug. Math. 33, 27–33 (1974).

25) G.W. Johnson, D.L. Skoug, A function space integral for a Banach space of functionals on Wiener space, Proc. Am. Math. Soc. 43, 141–148 (1974).

26) G.W. Johnson, D.L. Skoug, Cameron and Storvick's function space integral for certain Banach spaces of functionals, Preprint, Univ. Nebraska.

27) G.W. Johnson, D.L. Skoug, The Cameron-Storvick function space integral: The L_1 theory, Preprint, Univ. Nebraska.

28) G.W. Johnson, D.L. Skoug, The Cameron Storvick function space integral: An $L(L_p, L_p,)$ theory, Preprint, Univ. Nebraska.

29) R.H. Cameron, D.A. Storvick, An integral equation related to Schrödinger equation with an application to integration in function space, in Problems in Analysis, A Sympos. In Honor of S. Bochner, Edt. C. Gunning, Princeton Univ. Press, Princeton (1970), 175–193.

30) G.N. Gestrin, On Feynman integral, Izd. Kark. Univ., Theory of functions and funct. analys. and appl. 12, 69–81 (1970) (russ.).

11. 1) W.G. Faris, The product formula for semigroups defined by Friedrichs extensions, Pa J. Math. 22, 147–79 (1967).

2) W.G. Faris, Product formulas for perturbations of linear operators, J. Funct. Anal. 1, 93–108 (1967).

3) B. Simon, Quantum Mechanics for Hamiltonians defined as Quadratic Forms, Princeton Univ. Press (1971), pp. 50–53.

4) P.R. Chernoff, Product formulas, nonlinear semigroups and addition of unbounded operators, Mem. Am. Math. Soc. 140 (1974).

5) C.N. Friedman, Semigroup product formulas, compression and continuous observation in quantum mechanics, Indiana Univ. Math. J. 21, 1001–1011 (1972).

12. E.V. Markov, τ-smooth functionals, Trans. Moscow Math. Soc. 20, 1–40 (1969) (transl.).

13. K. Ito, Generalized uniform complex measures in the Hilbertian metric space with their application to the Feynman path integral, Proc. Fifth Berkeley Symposium on Mathematical Statistics and Probability, Univ. California Press, Berkeley (1967), Vol. II, part 1, pp. 145–161. See also K. Ito, Wiener integral and Feynman integral, Proc. Fourth Berkeley Symp. Math. Stat. and Prob., Univ. California Press, Berkeley (1961), Vol. 2, pp. 227–238.

14. 1) J. Tarski, Intégrale d' "histoire" pour le champ quantifié libre scalaire, Ann. Inst. Henri Poincaré, 15, 107–140 (1971).

2) J. Tarski, Mesures généralisées invariantes, Ann. Inst. Henri Poincaré, 17, 313–324 (1972). See also Ref. [6],5.

15. 1) C. Morette, De Witt, Feynman's path integral, Definition without limiting procedure, Commun. Math. Phys. 28, 47–67 (1972).

2) C. Morette, De Witt, Feynman path integrals, I. Linear and affine techniques, II. The Feynman – Green function, Commun. Math. Phys. 37, 63–81 (1974).

16. W. Garczynski, Quantum stochastic processes and the Feynman path integral for a single spinless particle, Repts. Mathem. Phys. 4, 21–46 (1973).

17. 1) E. Nelson, Derivation of the Schrödinger equation from Newtonian me-
        chanics, Phys. Rev. 150, 1079–1085 (1966).
     2) E. Nelson, Ref. [8].
     3) T.G. Dankel, Jr., Mechanics on manifolds and the incorporation of spin
        into Nelson's stochastic mechanics, Arch. Rat. Mech. Analysis 37, 192–221
        (1970).
     4) F. Guerra-P. Ruggiero, New interpretation of the Euclidean-Markow field
        in the framework of physical Minkowski space-time, Phys. Rev. Letts 31,
        1022–1025(1973).
     5) F. Guerra, On the connection between Euclidean-Markov field theory and
        stochastic quantization, to appear in Proc. Scuola Internazionale di Fisica
        "Enrico Fermi", Varenna (1973).
     6) F. Guerra, On stochastic field theory, Proc. II Aix-en-Provence Intern,
        Conf. Elem. Part. 1973, Journal de Phys. Suppl. T. 34, Fasc. 11–12, Col-
        loque, C - 1 - 95–98.
     7) S. Albeverio, R. Høegh-Krohn, A remark on the connection between sto-
        chastic mechanics and the heat equation, J. Math. Phys. 15, 1745–1748
        (1974).
18. 1) Gelfand-Yaglom, Ref. [5]
     2) Brush, Ref. [6], 3).
     3) C. Morette, Ref. [6], 4).
     4) C. Morette, De Witt, Ref. [14]
     5) G. Rosen, Ref. [7], 4).
     6) J. Tarski, Ref. [6], 5).
     7) N.N. Bogoliubov, D.V. Shirkov, Introduction to the theory of quantized
        fields, Interscience Publ., New York (1959), Ch. VII, p. 484.
     8) Yu. V. Novozhilov, A.V. Tulub, The Methods of Functionals in the Quan-
        tum Theory of Fields, Gordon and Breach, New York (1961).
     9) J. Rzewuski, Edt. Acta Univ. Wratisl. No 88, Functional methods in
        quantum field theory and statistical mechanics, (Karpacz, 1967), Wroclaw
        (1968)
    10) J. Rzewuski, Field Theory, Iliffe Books, London, PWN, Warsaw (1969).
    11) H.M. Fried, Functional Methods and Models in Quantum Field Theory,
        MIT Press, Cambridge (1972).
    12) E.S. Fradkin, U. Esposito, S. Termini, Functional techniques in physics,
        Rivista del Nuovo Cimento, Ser. I, 2, 498–560 (1970). Also Ref. [23] below.
19. 1) K.O. Friedrichs, Mathematical Aspects of the Quantum Theory of Fields,
        Interscience Publ., New York (1953)
     2) K.O. Friedrichs, Shapiro, Integration over Hilbert space and other exten-
        sions, Proc. Nat. Ac. Sci. 43, 336–338 (1957).
20. 1) M. Gelfand, A.M. Yaglom, Ref. [5].
     2) I.M. Gelfand, N.Ya. Vilenkin, Generalized Functions, Vol.4, Applications
        of harmonic analysis, Academic Press, New York (1964) (transl.)
21. 1) L. Gross, Harmonic Analysis on Hilbert space, Mem. Am. Math. Soc. No
        46 (1963), and references given therein.
     2) L. Gross, Classical analysis on a Hilbert space, in Ref. [23], 1) below.
22. 1) I. Segal, Tensor algebras over Hilbert spaces, I Trans. Am. Math. Soc. 81,
        106–134 (1956); II Ann. of Math. (2) 63, 160–175 (1956).
     2) E. Segal, Distributions in Hilbert space and canonical systems of operators,
        Trans. Am. Math. Soc. 88, 12–41 (1958).

3) I. Segal, Mathematical Problems of Relativistic Physics, AMS, Providence (1963), and References given therein.

4) I. Segal, Quantum fields and analysis in the solution manifolds of differential equations, in Ref. [23], below, pp. 129–153.

5) I. Segal, Algebraic integration theory, Bull. Am. Math. Soc. 71, 419–489 (1975).

23. 1) W. Ted Martin, I. Segal, Edts., Proc. Conference on Theory and Applications of Analysis in Function Spaces, MIT Press, Cambridge (1964).

2) R. Goodman, I. Segal, Edts., Mathematical Theory of Elementary Particles, MIT Press, Cambridge (1965).

3) Xia Dao, Xing, Measure and integration theory on infinite dimensional spaces, Academic Press, New York (1972).

24. 1) G. Velo, A. Wightman, Edts., Constructive Quantum Field Theory, Lecture Notes in Physics, 25, Springer, Berlin (1973).

2) B. Simon, The $P(\varphi)_2$ Euclidean (Quantum) Field Theory, Princeton Univ. Press, Princeton (1974).

25. 1) J. Glimm, A. Jaffe, in C. De Witt, R. Stora, Edts., Statistical Mechanics and Quantum Field Theory, Gordon and Breach, New York (1971), pp. 1–108.

2) J. Glimm, A. Jaffe, Boson quantum field models, in R.F. Streater, Edt., Mathematics of Contemporary Physics, Acad. Press London (1972), pp. 77–143.

3) I. Segal, Construction of non-linear local quantum processes, I, Ann of Math. (1) 92, 462–481 (1970), Erratum, ibid (2) 93, 597 (1971); II, Invent. Math. 14, 211–241 (1971).

26. J. Symanzik, Euclidean Quantum Field Theory, in E. Jost, Edt., Proc. Int. School of Physics "E. Fermi", Varenna, Local Quantum Theory, Academic Press, New York (1969), pp. 152–226, and references given therein.

27. E. Nelson, Probability theory and Euclidean field theory, in Ref. [24], 1, pp. 94–124.

28. S. Albeverio, R. Høegh-Krohn, The Wightman axioms and the mass gap for strong interactions of exponential type in two-dimensional space-time, J. Funct. Anal 16, 39–82 (1974).

29. 1) J. Glimm, A. Jaffe, Positivity of the φ_3^4 Hamiltonian, Fortschr. d. Phys. 21, 327 (1973).

2) J. Feldman, The $\lambda\varphi_3^4$ field theory in a finite volume, Commun. Math. Phys. 37, 93–120 (1974).

30. 1) S. Albeverio, R. Høegh-Krohn, Uniqueness of the physical vacuum and the Wightman functions in the infinite volume limit for some non polynomial interactions, Commun. Math. Phys. 30, 171–200 (1973).

2) S. Albeverio, R. Høegh-Krohn, The scattering matrix for some non polynomial interactions, I, Helv. Phys. Acta 46, 504–534 (1973); II, Helv. Phys. Acta 46, 535–545 (1973).

3) S. Albeverio, R. Høegh-Krohn, Asymptotic series for the scattering operator and asymptotic unitarity of the space cut-off interactions, Nuovo Cimento 18A, 285–370 (1973).

31. S.S. Schweber, On Feynman quantization, J. Math. Phys. 3, 831–842 (1962).

32. J. la Vita, J.P. Marchand, Edts., Scattering Theory in Mathematical Physics, D. Reidel Publ., Dordrecht (1974).

33. 1) D. Ruelle, Statistical Mechanics, Benjamin, New York (1969).

2) D.W. Robinson, The thermodynamic Pressure in Quantum Statistical Mechanics, Lecture Notes in Physics, 9, Springer (1971).

3) R. Høegh-Krohn, Relativistic quantum statistical mechanics in two-dimensional space-time, Commun. Math. Phys. 38, 195–224 (1974).

4) J. Ginibre, Some applications of functional integration in statistical mechanics, pp. 327–427 in the same bock as [25], 1).

5) S. Albeverio, R. Høegh-Krohn, Homogeneous random fields and statistical mechanics, J. Funct. Analys. 19, 242–272 (1975).

34. 1) D.W. Robinson, The ground state of the Bose gas, Commun. Math. Phys. 1, 159–174 (1965).

2) J. Manuceau, A. Verbeure, Quasi-free states of the C.C.R. algebra and Bogoliubov transformations, Commun. Math. Phys. 8, 293–302 (1968).

3) A. Van Daele, Quasi equivalence of quasi-free states on the Weyl algebra, Commun. Math. Phys. 22, 171-191 (1971).

35. F.A. Berezin, M.A. Subin, Symbols of operators and quantization, Colloquia Mathematica Societatis János Bolyai, 5. Hilbert space operators, Tihany (Hungary), 1970, pp. 21-52.

36. J. Tarski, Feynman-type integrals for spin and the functional approach to quantum field theory, Univ. Hamburg Preprint, June, 1974.

37. J. Eells, K.D. Elworthy, Wiener integration on certain manifolds, in Problems in non-linear analysis, C.I.M.E. IV (1970), pp. 67–94.

38. V.P. Maslov, Théorie des perturbations et méthóds asymptotiques, Dunod, Paris (1972) (transl.).

39. L.D. Faddeev, Symplectic structure and quantization of the Einstein gravitation theory, Actes Congrès Intern. Math., 1970 Tome 3, pp. 35–39, Gauthiers-Villars, Paris (1971).

40. R. Jost, The general theory of quantized fields, AMS, Providence (1965).
R. Streater, A. Wightman, PCT, Spin and Statistics, and all That, Benjamin, New York (1964).

41. S. Albeverio, R. Høegh-Krohn, Oscillatory integrals and the method of stationary phase in infinitely many dimensions, with applications to the classical limit of quantum mechanics I, Oslo University Preprint, Institute of Mathematics, September 1975.
Published in: Invent. Math. 40, no. 1, 59–106 (1977).

42. S. Albeverio, Lectures on Mathematical theory of Feynman path integrals, Preceedings of the XII-th Winter School of Theoretical Physics in Karpacz, Acta Universitatis Wratisläviensis.
Published in: "Probabilistic and functional methods in quantum field theory", Acta Universitatis Wratislaviensis, 85–139, Wroclaw (1976).

43. Séminaire Paul Krée, 1re année: 1974/75 Equations aux dérivées partielles en dimension infinie, Université Pierre et Marie Curie, Institut Henri Poincaré, Paris, 1975.

44. 1) J.S. Feldman, K. Osterwalder, The Wightman axioms and the mass gap for weakly counled $(\Phi^4)_3$ quantum field theories, in H. Araki Ed., International Symposium on Mathematical Problems in Theoretical Physics, Springer 1975.

2) J. Magnen, R. Sénéor, The infinite volume limit of the φ_3^4 model, Ann. Inst. H. Poincaré Sect. A (N.S.) 24, no. 2, 95–159 (1976).

45. S. Albeverio, Lectures on the construction of quantum fields with non polynomial interactions, Proceedings of the XII-th Vinter School of Theoretical Physics in Karpacz, Acta Universitatis Wratislaviensis. *Published in:* "Probabilistic and functional methods in quantum field theory", Acta Universitatis Wratislaviensis, 139–205, Wroclaw (1976).

46. K.R. Parthasaraty, Probability measures on metric spaces, Academic Press, New York (1967).

47. T. Kato, Perturbation Theory for Linear Operators, Springer (1966), Ch.X, §3, p. 527.

48. T. Kato, Some results on potential scattering, Proc. Intern. Conf. Functional Analysis and Related Topics, 1969, Univ. Tokyo Press (1970), pp. 206–215.

49. S.T. Kuroda, On the existence and unitary property of the scattering operator, Nuovo Cimento 12, 431–454 (1950).

50. R. Høegh-Krohn, Gentle perturbations by annihilation creation operators, Commun. Pure Appl. Math. 21, 343–357 (1968).

51. R. Høegh-Krohn, Partly gentle perturbation with application to perturbation by annihilation-creation operators, Proc. Nat. Ac. Sci. 58, 2187–2192 (1967).

52. K. Yosida, Functional Analysis, 2nd Ed., Springer, Berlin (1968).

53. 1) R. Høegh-Krohn, Infinite dimensional analysis with applications to self interacting boson fields in two space time dimensions, Proc. Aarhus Conf. Funct. Analys., Spring 1972.
    2) S. Albeverio, An introduction to some mathematical aspects of scattering theory in models of quantum fields, in Ref. [32], pp. 299–381.

54. R.F. Streater, Edt., Mathematics of Contemporary Physics, Academic Press, London (1972), particularly the articles by P.J. Bongaarts, Linear fields according to I.E. Segal, pp.187–208 and by B. Simon, Topics in functional analysis, 67–76.
    Also e.g. G.G. Emch, Algebraic methods in statistical mechanics and quantum field theory, Wiley, New York (1972).

55. M. Fierz, Über die Bedeutung der Funktion D_C in der Quantentheorie der Wellenfelder, Helv. Phys. Acta 23, 731–739 (1950).

56. F. Riesz, B. Sz.-Nagy, Functional analysis, Dover Publications, New York (1990).

References Added for the Second Edition

57. L. Accardi, Y. G. Lu, I. V. Volovich, Quantum theory and its stochastic limit. Springer-Verlag, Berlin, 2002.
58. L. Accardi, O. G. Smolyanov, Feynman formulas for evolution equations with Levy Laplacians on manifolds. Quantum probability and infinite dimensional analysis, 13–25, QP–PQ: Quantum Probab. White Noise Anal., 20, World Sci. Publ., Hackensack, NJ, 2007.
59. L. Accardi, I. V. Volovich, Feynman functional integrals and the stochastic limit. Stochastic analysis and mathematical physics (Via del Mar, 1996), 1–11, World Sci. Publ., River Edge, NJ, 1998.
60. S. Albeverio, Some recent developments and applications of path integrals. Path integrals from meV to MeV (Bielefeld, 1985), 3–32, Bielefeld Encount. Phys. Math., VII, World Sci. Publishing, Singapore, (1986).
61. S. Albeverio, T. Arede, The relation between quantum mechanics and classical mechanics: a survey of mathematical aspects. pp 37–76 in "Chaotic behavior in Quantum Systems, Theory and Applications", (Proc. Como 1983) Ed. G. Casati, Plenum Press, New York (1985).
62. S. Albeverio, Ph. Blanchard, R. Høegh-Krohn, Reduction of nonlinear problems to Schrödinger or heat equations: formation of Kepler orbits, singular solutions for hydrodynamical equations. Stochastic aspects of classical and quantum systems (Marseille, 1983), 189–206, Lecture Notes in Math., 1109, Springer, Berlin, 1985.
63. S. Albeverio, Ph. Blanchard, Ph. Combe, R. Høegh-Krohn, M. Sirugue, Local relativistic invariant flows for quantum fields, Comm. Math. Phys. 90 (3), 329–351 (1983).
64. S. Albeverio, Ph. Blanchard, R. Høegh-Krohn, Stationary phase for the Feynman integral and the trace formula, in: Functional integration, Plenum, New York (1980)
65. S. Albeverio, Ph. Blanchard, R. Høegh-Krohn, Feynman path integrals, the Poisson formula and the theta function for the Schrödinger operators, in: Trends in Applications of Pure Mathematics to Mechanics, Vol III. Pitman, Boston, 1–21 (1981).
66. S. Albeverio, Ph. Blanchard, R. Høegh-Krohn, Feynman path integrals and the trace formula for the Schrödinger operators, Comm. Math. Phys. 83 n.1, 49–76 (1982).

67. S. Albeverio, A.M. Boutet de Monvel-Berthier, Z. Brzeźniak, The trace formula for Schrödinger operators from infinite dimensional oscillatory integrals, Math. Nachr. 182, 21–65 (1996).

68. S. Albeverio, A. Boutet de Monvel-Berthier, Z. Brzeźniak, Stationary phase method in infinite dimensions by finite-dimensional approximations: applications to the Schrödinger equation. Potential Anal. 4 (1995), no. 5, 469–502.

69. S. Albeverio, Z. Brzeźniak, Finite-dimensional approximation approach to oscillatory integrals and stationary phase in infinite dimensions, J. Funct. Anal. 113 no. 1, 177–244 (1993).

70. S. Albeverio, Z. Brzeźniak, Oscillatory integrals on Hilbert spaces and Schrödinger equation with magnetic fields, J. Math. Phys. 36 (5), 2135–2156 (1995).

71. S. Albeverio, Z. Brzeźniak, Feynman path integrals as infinite-dimensional oscillatory integrals: some new developments. Acta Appl. Math 35, 5–27 (1994).

72. S. Albeverio, Z. Brzeźniak, The detailed dependence of solution of Schrödinger equation on Planck's constant and the trace formula, in "Stochastic processes in physics and geometry", Eds. S. Albeverio, U. Cattaneo, D. Merlini. World Scientific, Singapore, pp 1–13 (1995).

73. S. Albeverio, Z. Brzeźniak, Z. Haba, On the Schrödinger equation with potentials which are Laplace transform of measures, Potential Anal. 9 (1), 65–82 (1998).

74. S. Albeverio, L. Cattaneo, S. Mazzucchi, L. Di Persio, A rigorous approach to the Feynman-Vernon influence functional and its applications. I. J. Math. Phys. 48 (2007), no. 10, 102–109, 22 pp.

75. S. Albeverio, Ph. Combe, R. Høegh-Krohn, G. Rideau, M. Sirugue, M. Sirugue-Collin, R. Stora, Mathematical problems in Feynman path integrals, Proceedings of the Marseille International Colloquium, May 1978, Lecture Notes in Physics 106, Springer (1979).

76. S. Albeverio, L. Di Persio, S. Mazzucchi, Semiclassical asymptotic expansion for the Schrödinger equation with polynomially growing potentials, In preparation.

77. S. Albeverio, S. Dobrokhotov, M. Poteryakhin, On quasimodes of small diffusion operators corresponding to stable invariant tori with nonregular neighborhoods. Asymptot. Anal. 43 (2005), no. 3, 171–203.

78. S. Albeverio, J.E. Fenstad, R. Høegh-Krohn, T. Lindstrøm, Non Standard Methods in Stochastic Analysis and Mathematical Physics, *Pure and Applied Mathematics* 122, Academic Press, Inc., Orlando, FL, (1986), 2nd edition, in preparation.

79. S. Albeverio, B. Ferrario, Some methods of infinite dimensional analysis in hydrodynamics: an introduction. Lectures from the Summer School held in Cetraro, August 29–September 3 2005. Fondazione C.I.M.E. [C.I.M.E. Foundation], Springer, Berlin (2008).

80. S. Albeverio, G. Guatteri, S. Mazzucchi, Phase space Feynman path integrals, J. Math. Phys. 43, 2847–2857 (2002).

81. S. Albeverio, G. Guatteri, S. Mazzucchi, Representation of the Belavkin equation via Feynman path integrals, Probab. Theory Relat. Fields 125, 365–380 (2003).

82. S. Albeverio, G. Guatteri, S. Mazzucchi, Representation of the Belavkin equation via phase space Feynman path integrals, Infin. Dimens. Anal. Quantum Probab. Relat. Top. 7, no. 4, 507–526 (2004).

83. S. Albeverio, A. Hahn, A. Sengupta, Rigorous Feynman path integrals, with applications to quantum theory, gauge fields, and topological invariants, Stochastic analysis and mathematical physics (SAMP/ANESTOC 2002), 1–60, World Sci. Publishing, River Edge, NJ, (2004).

84. S. Albeverio, A. Hilbert, V. Kolokoltsov, Transience of stochastically perturbed classical Hamiltonian systems and random wave operators. Stochastics Stochastics Rep. 60, no. 1-2, 41–55 (1997).

85. S. Albeverio, A. Hilbert, V. Kolokoltsov, Estimates uniform in time for the transition probability of diffusions with small drift and for stochastically perturbed Newton equations. J. Theoret. Probab. 12, no. 2, 293–300 (1999).

86. S. Albeverio, A. Hilbert, V. Kolokoltsov, Sur le comportement asymptotique du noyau associé à une diffusion dégénérée. (French) [Asymptotic behavior of the kernel associated with a degenerate diffusion] C. R. Math. Acad. Sci. Soc. R. Can. 22, no. 4, 151–159 (2000).

87. S. Albeverio, R. Høegh-Krohn, Oscillatory integrals and the method of stationary phase in infinitely many dimensions, with applications to the classical limit of quantum mechanics, Invent. Math. 40 n.1, 59–106 (1977).

88. S. Albeverio, R. Høegh-Krohn, Uniqueness and the global Markov property for euclidean fields. The case of trigonometric interaction, Commun. Math. Phys. 68 n.2, 95–128 (1979).

89. S. Albeverio, J.W. Johnson, Z.M. Ma, The analytic operator-value Feynman integral via additive functionals of Brownian motion, Acta Appl. Math. 42 n.3, 267–295 (1996).

90. S. Albeverio, J. Jost, S. Paycha, S. Scarlatti, A mathematical introduction to string theory. Variational problems, geometric and probabilistic methods. London Mathematical Society Lecture Note Series, 225. Cambridge University Press, Cambridge, 1997.

91. S. Albeverio, W. Karwowski, K. Yasuda, Trace formula for p-adics. Acta Appl. Math. 71, no. 1, 31–48 (2002).

92. S. Albeverio, A. Khrennikov, O. Smolyanov, The probabilistic Feynman-Kac formula for an infinite-dimensional Schrödinger equation with exponential and singular potentials, Potential Anal. 11, no. 2, 157–181 (1999).

93. S. Albeverio, A. Khrennikov, O. Smolyanov, Representation of solutions of Liouville equations for generalized Hamiltonian systems by functional integrals. (Russian) Dokl. Akad. Nauk 381, no. 2, 155–159 (2001).

94. S. Albeverio, A. Khrennikov, O. Smolyanov. Solution of Schrödinger equations for finite-dimensional Hamilton-Dirac systems with polynomial Hamiltonians, Dokl. Akad. Nauk. 397 no. 2, 151–154 (2004).

95. S. Albeverio, V. N. Kolokoltsov, The rate of escape for some Gaussian processes and the scattering theory for their small perturbations. Stochastic Process. Appl. 67 (1997), no. 2, 139–159.

96. S. Albeverio, V. N. Kolokoltsov, O. G. Smolyanov, Représentation des solutions de l'équation de Belavkin pour la mesure quantique par une version rigoureuse de la formule d'intégration fonctionnelle de Menski. C. R. Acad. Sci. Paris Sér. I Math. 323 (6), 661–664 (1996).

97. S. Albeverio, V. N. Kolokoltsov, O. G. Smolyanov, Continuous quantum measurement: local and global approaches, Rev. Math. Phys. 9 n.8, 907–920 (1997).

98. S. Albeverio, Y. Kondratiev, Yu. Kozitsky, M. Röckner, Euclidean Gibbs states of quantum lattice systems, Rev. Math. Phys. 14, no. 12, 1335–1401 (2002).

99. S. Albeverio, Y. Kondratiev, Yu. Kozitsky, M. Röckner, Euclidean Gibbs states of quantum anharmonic crystals, book in preparation.

100. S. Albeverio, Y. Kondratiev, T. Pasurek, M. Röckner, Euclidean Gibbs measures of quantum cristals: existence, uniqueness and a priori estimates, Interacting stochastic systems, Springer, Berlin, 29–54 (2005).

101. S. Albeverio, Y. Kondratiev, T. Pasurek, M. Röckner, Euclidean Gibbs measures on loop lattices: existence and a priori estimates, Ann. Probab. 32 (1A), 153–190 (2004).

102. S. Albeverio, S. Liang, Asymptotic expansions for the Laplace approximations of sums of Banach space-valued random variables. Ann. Probab. 33, no. 1, 300–336 (2005).

103. S. Albeverio, S. Mazzucchi, Generalized Fresnel Integrals, Bull. Sci. Math. 129, no. 1, 1–23 (2005).

104. S. Albeverio, S. Mazzucchi, Generalized infinite-dimensional Fresnel Integrals, C. R. Acad. Sci. Paris 338 n.3, 255–259 (2004).

105. S. Albeverio, S. Mazzucchi, Feynman path integrals for polynomially growing potentials, J. Funct. Anal. 221 no.1, 83–121 (2005).

106. S. Albeverio, S. Mazzucchi, Some New Developments in the Theory of Path Integrals, with Applications to Quantum Theory, J. Stat. Phys. 115 n.112, 191–215 (2004).

107. S. Albeverio, S. Mazzucchi, Feynman Path integrals for time-dependent potentials, in : "Stochastic Partial Differential Equations and Applications -VII", G. Da Prato and L. Tubaro eds, Lecture Notes in Pure and Applied Mathematics, vol. 245, Taylor & Francis, (2005), pp 7-20.

108. S. Albeverio, S. Mazzucchi, Feynman path integrals for the time dependent quartic oscillator. C. R. Acad. Sci. Paris 341, no. 10, 647–650. (2005).

109. S. Albeverio, S. Mazzucchi, The time dependent quartic oscillator - a Feynman path integral approach. J. Funct. Anal. 238, no. 2, 471–488 (2006).

110. S. Albeverio, S. Mazzucchi, The trace formula for the heat semigroup with polynomial potential, SFB-611-Preprint no. 332, Bonn (2007).

111. S. Albeverio, S. Mazzucchi, Theory and applications of infinite dimensional oscillatory integrals. pp 73–92 in : "Stochastic Analysis and Applications", Proceedings of the Abel Symposium 2005 in honor of Prof. Kiyosi Ito, Springer (2007).

112. S. Albeverio and I. Mitoma. Asymptotic Expansion of Perturbative Chern-Simons Theory via Wiener Space. SFB-611-Preprint no. 322, Bonn (2007). to appear in Bull. Sci. Math. (2007).

113. S. Albeverio, J. Rezende, J.-C. Zambrini, Probability and quantum symmetries. II. The theorem of Noether in quantum mechanics. J. Math. Phys. 47 (2006), no. 6, 062107, 61 pp.

114. S. Albeverio, H. Röckle, V. Steblovskaya. Asymptotic expansions for Ornstein-Uhlenbeck semigroups perturbed by potentials over Banach spaces. Stochastics Stochastics Rep. 69, n.3-4, 195-238 (2000).

115. S. Albeverio, J. Schäfer, Abelian Chern-Simons theory and linking numbers via oscillatory integrals, J. Math. Phys. 36, 2157–2169 (1995).

116. S. Albeverio, A. Sengupta, A mathematical construction of the non-Abelian Chern-Simons functional integral, Commun. Math. Phys. 186, 563–579 (1997).

117. S. Albeverio, A. Sengupta, The Chern-Simons functional integral as an infinite-dimensional distribution. Proceedings of the Second World Congress of Nonlinear Analysts, Part 1 (Athens, 1996). Nonlinear Anal. 30 (1997), no. 1, 329–335

118. S. Albeverio, O. G. Smolyanov, E. T. Shavgulidze, Solution of Schrödinger equations for finite-dimensional Hamilton-Dirac systems with polynomial Hamiltonians. (Russian) Dokl. Akad. Nauk 397, no. 2, 151–154 (2004).

119. S. Albeverio, V. Steblovskaya. Asymptotics of infinite-dimensional integrals with respect to smooth measures I. Infin. Dimens. Anal. Quantum Probab. Relat. Top. 2, n.4, 529-556 (1999).

120. S. Albeverio, Wiener and Feynman Path Integrals and Their Applications. Proc. Symp. Appl. Math. 52, 163–194 (1997); an extended version of the references is in Note di Matematica e Fisica, Anno 9, Vol. 8, Ed. CERFIM, Locarno.

121. S. Albeverio, B. Rüdiger, Stochastic integrals and the Lévy-Ito decomposition theorem on separable Banach Spaces. Stoch. Anal. Appl. 23, no. 2, 217–253 (2005).

122. D. Applebaum, Lévy processes and stochastic calculus. Cambridge Studies in Advanced Mathematics, 93. Cambridge University Press, Cambridge, (2004).

123. V. I. Arnold, Integrals of quickly oscillating functions and singularities of projections of Lagrange manifolds, Funct. Analys. and its Appl. 6, 222–224 (1972).

124. V. I. Arnold, V. V. Goryunov, O. V. Lyashko, V. A. Vassiliev, Singularity theory. I. Translated from the 1988 Russian original by A. Iacob. Reprint of the original English edition from the series Encyclopaedia of Mathematical Sciences [Dynamical systems]. VI, Encyclopaedia Math. Sci., 6, Springer, Berlin, 1993; Springer-Verlag, Berlin, 1998.

125. A.M. Arthurs, Ed., Functional integration and its applications, Clarendon Press, Oxford (1975).

126. N. Asai, I. Kubo, H.H. Kuo, Characterization of test functions in CKS-space. Mathematical physics and stochastic analysis (Lisbon, 1998), 68–78, World Sci. Publ., River Edge, NJ, (2000).

127. M. Atiyah, The Geometry and Physics of Knots, Cambridge University Press, Cambridge (1990).

128. R. Azencott, Formule de Taylor stochastique et développement asymptotique d'intégrales de Feynman. (French) [Stochastic Taylor formula and asymptotic expansion of Feynman integrals]. Seminar on Probability, XVI, Supplement, pp. 237–285, Lecture Notes in Math., 921, Springer, Berlin-New York, (1982).

129. R. Azencott, H. Doss, L'équation de Schrödinger quand h tend vers zéro: une approche probabiliste. (French) [The Schrödinger equation as h tends to zero: a probabilistic approach], Stochastic aspects of classical and quantum systems (Marseille, 1983), 1–17, Lecture Notes in Math. 1109, Springer, Berlin, (1985).

130. W. Balser, Formal power series and linear systems of meromorphic ordinary differential equations. Universitext. Springer-Verlag, New York, (2000).

131. A. Barchielli, On the quantum theory of measurements continuous in time, Rep. Math. Phys. 33, n 1-2, 21–34 (1993).

132. A. Barchielli, V.P. Belavkin, Measurements continuous in time and a posteriori states in quantum mechanics, J. Phys. A 24, n 7, 1495–1514 (1991).

133. D. Bauch, The path integral for a particle moving in a δ-function potential. Nuovo Cimento B (11) 85 (1985), no. 1, 118–124.

134. V.P. Belavkin, A new wave equation for a continuous nondemolition measurement, Phys. Lett. A140, no.7/8, 355–358 (1989).

135. V. P. Belavkin, V.N. Kolokoltsov, Quasiclassical asymptotics of quantum stochastic equations. (Russian) Teoret. Mat. Fiz. 89 (1991), no. 2, 163–177; translation in Theoret. and Math. Phys. 89 (1991), no. 2, 1127–1138 (1992).

136. V. P. Belavkin, V.N. Kolokoltsov, Stochastic evolution as a quasiclassical limit of a boundary value problem for Schrödinger equations. Infin. Dimens. Anal. Quantum Probab. Relat. Top. 5, no. 1, 61–91 (2002).

137. V. V. Belokurov, Yu. P. Solovev, E. T. Shavgulidze, New perturbation theory for quantum field theory: convergent series instead of asymptotic expansions. Mod. Phys. Let. A 19, no. 39, 3033–3041 (1995).

138. V. V. Belokurov, Yu. P. Solovev, E. T. Shavgulidze, A method for the approximate calculation of path integrals using perturbation theory with convergent series. I. (Russian) Teoret. Mat. Fiz. 109 (1996), no. 1, 51–59; translation in Theoret. and Math. Phys. 109 (1996), no. 1, 1287–1293 (1997)

139. V. V. Belokurov, Yu. P. Solovev, E. T. Shavgulidze, A method for the approximate calculation of path integrals using perturbation theory with convergent series. II. Euclidean quantum field theory. (Russian) Teoret. Mat. Fiz. 109 (1996), no. 1, 60–69; translation in Theoret. and Math. Phys. 109 (1996), no. 1, 1294–1301 (1997)

140. V. V. Belokurov, Yu. P. Solovev, E. T. Shavgulidze, Perturbation theory with convergent series for functional integrals with respect to the Feynman measure. (Russian) Uspekhi Mat. Nauk 52, no. 2(314), 155–156 (1997); translation in Russian Math. Surveys 52, no. 2, 392–393 (1997).

141. V. V. Belov, S. Yu. Dobrokhotov, S. O. Sinitsyn, Asymptotic solutions of the Schrödinger equation in thin tubes. Proc. Steklov Inst. Math., Asymptotic Expansions. Approximation Theory. Topology, suppl. 1, S13–S23 (2003).

142. G. Ben Arous, F. Castell, A probabilistic approach to semi-classical approximations, J. Funct. Anal. 137, no. 1, 243–280 (1996).

143. G. Ben Arous, A. Rouault, Laplace asymptotics for reaction-diffusion equations. Probab. Theory Related Fields 97, no. 1-2, 259–285 (1993).

144. F. Benatti, G.C. Ghirardi, A. Rimini, T. Weber, Quantum mechanics with spontaneous localization and the quantum theory of measurement, Nuovo Cimento B (11) 100, 27–41 (1987).

145. C.M. Bender, S.A. Orszag, Advanced mathematical methods for scientists and engineers. I. Asymptotic methods and perturbation theory. Reprint of the 1978 original. Springer-Verlag, New York, (1999).

146. F.A. Berezin, M.A. Shubin, The Schrödinger Equation, Kluwer, Dordrecht, (1991).

147. C.C. Bernido, Exact path summation for relativistic particles in flat and curved spacetime. Path integrals from peV to TeV (Florence, 1998), 275–278, World Sci. Publ., River Edge, NJ, (1999).

148. C.C. Bernido, G. Aguarte, Summation over histories for a particle in spherical orbit around a black hole. Phys. Rev. D (3) 56, no. 4, 2445–2448 (1997).

149. C.C. Bernido, M.V. Carpio-Bernido, Path integrals for boundaries and topological constraints: a white noise functional approach. J. Math. Phys. 43, no. 4, 1728–1736 (2002).

150. C.C. Bernido, M.V. Carpio-Bernido, Entanglement probabilities of polymers: a white noise functional approach. J. Phys. A 36, no. 15, 4247–4257 (2003).

151. C.C. Bernido, A. Inomata, Path integrals with a periodic constraint: the Aharonov-Bohm effect. J. Math. Phys. 22, no. 4, 715–718 (1981).

152. C. Berns, Asymptotische Entwicktlungen von Laplace Integralen mit Phasenfunktionen vierten Graden und Borel-Summierbarkeit. Dipl. Thesis, Bonn (2007).

153. M.V. Berry, Asymptotics of evanescence. J. Modern Opt. 48, no. 10, 1535–1541 (2001).

154. M. V. Berry, Making light of mathematics. Bull. Amer. Math. Soc. (N.S.) 40, no. 2, 229–237 (2003).

155. G. D. Birkhoff, Quantum mechanics and asymptotic series. Bull. AMS 39, 681–700 (1933).

156. K. Bitar, N. N. Khuri, H. C. Ren, Path integrals as discrete sums. Phys. Rev. Lett. 67, no. 7, 781–784 (1991).

157. Ph. Blanchard, Ph. Combe, M. Sirugue, M. Sirugue-Collin, Jump processes: an introduction and some applications in quantum theories. Rend. Circ. Mat. Palermo Ser. II, no 17, 47-104 (1987).

158. A. Bouzouina, D. Robert, Uniform semiclassical estimates for the propagation of quantum observables. Duke Math. J. 111, no. 2, 223–252 (2002).

159. H. Breuer, F. Petruccione, The theory of quantum open systems. Oxford University Press, New York, 2002.

160. B. Broda, A path-integral approach to polynomial invariants of links. Topology and physics. J. Math. Phys. 35, no. 10, 5314–5320 (1994).

161. L. S. Brown, Quantum field theory. Cambridge University Press, Cambridge, (1992).

162. J. Brüning, S. Yu. Dobrokhotov, K. V. Pankrashkin, The spectral asymptotics of the two-dimensional Schrödinger operator with a strong magnetic field. I. Russ. J. Math. Phys. 9, no. 1, 14–49 (2002).

163. A. Budhiraja, G. Kallianpur, The Feynman-Stratonovich semigroup and Stratonovich integral expansions in nonlinear filtering. Appl. Math. Optim. 35, no. 1, 91–116 (1997).

164. A. A. Bytsenko, G. Cognola, E. Elizalde, V. Moretti, S. Zerbini, Analytic aspects of quantum fields. World Scientific Publishing Co., Inc., River Edge, NJ, (2003).

165. R.H. Cameron, A family of integrals serving to connect the Wiener and Feynman integrals, J. Math. and Phys. 39, 126–140 (1960).

166. R.H. Cameron, D.A. Storvick, An operator-valued function-space integral applied to integrals of functions of class L_1, Proc. London Math. Soc. 3rd series, 27, 345–360 (1973).

167. R.H. Cameron, D.A. Storvick, Some Banach algebras of analytic Feynman integrable functionals, Analytic functions, Kozubnik, 1979. Lecture Notes in Mathematics, n 798, Springer-Verlag, Berlin, p. 18-67 (1980).

168. R.H. Cameron, D.A. Storvick, A simple definition of the Feynman integral with applications functionals, Memoirs of the American Mathematical Society n. 288, 1–46 (1983).

169. P. Cartier, C. DeWitt-Morette, Functional integration, J. Math. Phys. 41, no. 6, 4154–4187 (2000).

170. P. Cartier, C. DeWitt-Morette, A rigorous mathematical foundation of functional integration. NATO Adv. Sci. Inst. Ser. B Phys., 361, Functional integration (Cargse, 1996), 1–50, Plenum, New York, (1997).

171. P. Cartier, C. DeWitt-Morette, A new perspective on functional integration. J. Math. Phys. 36, no. 5, 2237–2312 (1995).

172. P. Cartier, C. DeWitt-Morette, Functional integration: action and symmetries. Appendix D contributed by Alexander Wurm. Cambridge Monographs on Mathematical Physics. Cambridge University Press, Cambridge, (2006).

173. E. Combet, Intégrales Exponentielles, Lecture Notes in Mathematics 937, Springer, Berlin, (1982).
174. S.-J. Chang, Introduction to quantum field theory. World scientific, Singapore (1990).
175. D. M. Chung, Conditional analytic Feynman integrals on Wiener spaces, Proc. AMS 112, 479-488 (1991).
176. M. Cunha, C. Drumond, P. Leukert, J. L. Silva, W. Westerkamp, The Feynman integrand for the perturbed harmonic oscillator as a Hida distribution. Ann. Physik (8) 4, no. 1, 53–67 (1995).
177. K. S. Chang, K. P. Hong, K. S. Ryu, Feynman's operational calculus for a sequential operator-valued function space integral. Acta Math. Hungar. 91, no. 1-2, 9–25 (2001).
178. K. S. Chang, J. A. Lim, K. S. Ryu, Stability theorem for the Feynman integral via time continuation. Rocky Mountain J. Math. 29, no. 4, 1209–1224 (1999).
179. L.K. Chung, J.C. Zambrini. Introduction to random time and quantum randomness. *Monographs of the Portuguese Mathematical Society* McGraw-Hill, Lisbon, (2001).
180. Y. Colin de Verdière, Paramétrix de l'équation des ondes et intégrales sur l'espace des chemins. (French) Séminaire Goulaouic-Lions-Schwartz (1974–1975), Équations aux dérivées partielles linéaires et non linéaires, Exp. No. 20, 13 pp. Centre Math., École Polytech., Paris, (1975).
181. Y. Colin de Verdière, Déterminants et intégrales de Fresnel. (French) [Determinants and Fresnel integrals] Symposium à la Mémoire de F. Jaeger (Grenoble, 1998). Ann. Inst. Fourier (Grenoble) 49, no. 3, 861–881 (1999).
182. Y. Colin de Verdière, Singular Lagrangian manifolds and semiclassical analysis. Duke Math. J. 116, no. 2, 263–298 (2003).
183. Ph. Combe, R. Høegh-Krohn, R. Rodriguez, M. Sirugue, M. Sirugue-Collin, Generalized Poisson processes in quantum mechanics and field theory, Phys. Rep. 77, 221–233 (1981).
184. Ph. Combe, R. Høegh-Krohn, R. Rodriguez, M. Sirugue, M. Sirugue-Collin, Poisson Processes on Groups and Feynman Path Integrals, Commun. Math. Phys. 77, 269–288 (1980).
185. Ph. Combe, R. Høegh-Krohn, R. Rodriguez, M. Sirugue, M. Sirugue-Collin, Feynman path integrals and Poisson processes with piecewise classical paths, J. Math. Phys. 23, 405–411 (1982).
186. J.M. Combes, Recent developments in quantum scattering theory, Mathematical problems in theoretical physics (Proc. Internat. Conf. Math. Phys., Lausanne, 1979), 1–24, Lecture Notes in Phys., 116, Springer, Berlin-New York, (1980).
187. J.M. Combes, Scattering theory in quantum mechanics and asymptotic completeness, Mathematical problems in theoretical physics (Proc. Internat. Conf., Univ. Rome, Rome, 1977), 183–204, Lecture Notes in Phys., 80, Springer, Berlin-New York, (1978).
188. Ph. Combe, R. Rodriguez, On the cylindrical approximation of the Feynman path integral. Rep. Math. Phys. 13, no. 2, 279–294 (1978).
189. M. Combescure, D. Robert, Semiclassical results in the linear response theory. Ann. Physics 305, no. 1, 45–59 (2003).
190. M. Cowling, S. Disney, G. Mauceri, D. Müller, Damping oscillatory integrals. Invent. Math. 101, no. 2, 237–260 (1990).

191. R. E. Crandall, Combinatorial approach to Feynman path integration. J. Phys. A 26, no. 14, 3627–3648 (1993).
192. A. Cruzeiro, J.C. Zambrini, Feynman's functional calculus and stochastic calculus of variations. Stochastic analysis and applications (Lisbon, 1989), 82–95, Progr. Probab., 26, Birkhuser Boston, Boston, MA, (1991).
193. H. L. Cycon, R. G. Froese, W. Kirsch, B. Simon, Schrödinger operators with application to quantum mechanics and global geometry. Texts and Monographs in Physics. Springer Study Edition. Springer-Verlag, Berlin, 1987.
194. Ju. L. Daleckiĭ, Functional integrals associated with operator evolution equations. Uspehi Mat. Nauk 17 no. 5 , 3–115 (1962).
195. Ju. L. Daleckiĭ, V. Ju. Krylov, R. A. Minlos, V. N. Sudakov, Functional integration and measures in function spaces. (Russian). Proc. Fourth All-Union Math. Congr. (Leningrad, 1961) (Russian), Vol. II pp. 282–292 Izdat. "Nauka", Leningrad (1964).
196. I. Daubechies, J. R. Klauder, Constructing measures for path integrals. J. Math. Phys. 23, no. 10, 1806–1822 (1982).
197. I. Daubechies, J. R. Klauder, Measures for more quadratic path integrals. Lett. Math. Phys. 7, no. 3, 229–234 (1983).
198. I. Daubechies, J. R. Klauder, Quantum-mechanical path integrals with Wiener measure for all polynomial Hamiltonians, Phys. Rev. Lett. 52, no. 14, 1161–1164 (1984).
199. I. Daubechies, J. R. Klauder, Quantum-mechanical path integrals with Wiener measure for all polynomial Hamiltonians II, J. Math. Phys. 26, 2239–2256 (1985).
200. I. Daubechies, J. R. Klauder, T. Paul, Wiener measures for path integrals with affine kinematic variables. J. Math. Phys. 28, no. 1, 85–102 (1987).
201. E. B. Davies, Quantum Theory of Open Systems, Academic Press, London, (1976).
202. I. M. Davies, Laplace asymptotic expansions for Gaussian functional integrals. Electron. J. Probab. 3, no. 13, 1–19 (1998).
203. I. M. Davies, A. Truman, Laplace expansions of conditional Wiener integrals and applications to quantum physics. Stochastic processes in quantum theory and statistical physics (Marseille, 1981), 40–55, Lecture Notes in Phys., 173, Springer, Berlin, (1982).
204. I. M. Davies, A. Truman, Laplace asymptotic expansions of conditional Wiener integrals and generalized Mehler kernel formulas. J. Math. Phys. 23, no. 11, 2059–2070 (1982).
205. I. M. Davies, A. Truman, On the Laplace asymptotic expansion of conditional Wiener integrals and the Bender-Wu formula for x^{2N}-anharmonic oscillators. J. Math. Phys. 24 , no. 2, 255–266 (1983).
206. M. De Faria, J. Potthoff, L. Streit, The Feynman integrand as a Hida distribution. J. Math. Phys. 32, 2123–2127 (1991).
207. M. De Faria, M. J. Oliveira, L. Streit, Feynman integrals for nonsmooth and rapidly growing potentials. J. Math. Phys. 46, no. 6,063505, 14 pp, (2005).
208. C. DeWitt-Morette, Feynman's path integral. Definition without limiting procedure. Comm. Math. Phys. 28, 47–67 (1972).
209. C. DeWitt-Morette, Feynman path integrals: from the prodistribution definition to the calculation of glory scattering. Stochastic methods and computer techniques in quantum dynamics (Schladming, 1984), 101–170, Acta Phys. Austriaca Suppl., XXVI, Springer, Vienna, (1984).

210. C. DeWitt-Morette, A. Maheshwari, Feynman path integration, Proc. XII-th Winter School of Theor. Physics, Karpacz, Acta Univ. Wratislaviensis (1976).
211. C. DeWitt-Morette, A. Maheshwari, B. Nelson, Path integration in phase space. General Relativity and Gravitation 8, no. 8, 581–593 (1977).
212. L. Diosi, Continuous quantum measurements and Ito formalism, Phys. Lett. A 129 (8), 419–223 (1988).
213. L. Di Persio, Asymptotic expansion of integrals in statistical mechanics and quantum theory, PhD thesis, Trento (2006).
214. I. Dobrakov, Feynman type integrals as multilinear integrals. I. Measure theory (Oberwolfach, 1990). Rend. Circ. Mat. Palermo (2) Suppl. No. 28, 169–180 (1992).
215. S. Y. Dobrokhotov, V. N. Kolokoltsov, On the amplitude of the splitting of lower energy levels of the Schrödinger operator with two symmetric wells. (Russian) Teoret. Mat. Fiz. 94 (1993), no. 3, 426–434; translation in Theoret. and Math. Phys. 94, no. 3, 300–305 (1993).
216. S. Y. Dobrokhotov, V. N. Kolokoltsov, The double-well splitting of the low energy levels for the Schrödinger operator of discrete φ^4-models on tori. J. Math. Phys. 36, no. 3, 1038–1053 (1995).
217. H. Doss, Sur une Résolution Stochastique de l'Equation de Schrödinger à Coefficients Analytiques, Commun. Math. Phys. 73, 247–264 (1980).
218. A. Doumeki, T. Ichinose, H. Tamura, Error bounds on exponential product formulas for Schrödinger operators. J. Math. Soc. Japan 50, no. 2, 359–377 (1998).
219. T. Dreyfus, H. Dym, Product formulas for the eigenvalues of a class of boundary value problems. Duke Math. J. 45, no. 1, 15–37 (1978).
220. J.J. Duistermaat, Oscillatory integrals, Lagrange immersions and unfoldings of singularities, Comm. Pure Appl. Math. 27, 207–281 (1974).
221. R. S. Egikyan, D. V. Ktitarev, The Feynman formula in phase space for systems of pseudodifferential equations with analytic symbols. Math. Notes 51, no. 5-6, 453–457 (1992).
222. I. Ekeland, Convexity methods in Hamiltonian mechanics, Springer Verlag, Berlin, (1990).
223. D. Elworthy, A. Truman, Feynman maps, Cameron-Martin formulae and anharmonic oscillators, Ann. Inst. H. Poincaré Phys. Théor. 41(2), 115–142 (1984).
224. D. Elworthy, A. Truman, The diffusion equation and classical mechanics: an elementary formula. Stochastic processes in quantum theory and statistical physics (Marseille, 1981), 136–146, Lecture Notes in Phys., 173, Springer, Berlin, (1982).
225. D. Elworthy, A. Truman, K. Watling, The semiclassical expansion for a charged particle on a curved space background. J. Math. Phys. 26, no. 5, 984–990 (1985).
226. A. Erdélyi, Asymptotic expansions, Dover Publications, Inc., New York, (1956).
227. R. Estrada, R. P. Kanwal, Distributional approach to asymptotics. Theory and applications. Second edition. Birkhäuser Advanced Texts: Basler Lehrbücher. [Birkhäuser Advanced Texts: Basel Textbooks] Birkhäuser Boston, Inc., Boston, MA, (2002).
228. P. Exner, Open quantum systems and Feynman integrals, Czecholslovak J. Phys. B 32, no 6, 628–632 (1982).

229. P. Exner, Complex potentials and rigorous Feynman integrals, Reidel, Dordrecht (1985).

230. M. V. Fedoryuk, Asymptotic analysis. Linear ordinary differential equations. Translated from the Russian by Andrew Rodick. Springer-Verlag, Berlin, (1993).

231. M. V. Fedoryuk, Asymptotic Methods in Analysis, in Enc. Math. Sci. 13, Anal. 1, Part II, Springer, Berlin (1989).

232. R. Feynman, Space-time approach to non-relativistic quantum mechanics, Rev. Mod. Phys. 20, 367–387 (1948).

233. R. Figari, R. Hoegh-Krohn, C.R. Nappi, Interacting relativistic boson fields in the de Sitter universe with two space-time dimensions. Comm. Math. Phys. 44 no. 3, 265–278 (1975).

234. P. Florchinger, R. Léandre, Décroissance non exponentielle du noyau de la chaleur. (French) [Nonexponential decay of the heat kernel] Probab. Theory Related Fields 95, no. 2, 237–262 (1993).

235. E. Franchini, M. Maioli, Feynman integrals with point interactions. Comput. Math. Appl. 46, no. 5-6, 685–694 (2003).

236. C. N. Friedman, Semigroup product formulas, compressions, and continual observations in quantum mechanics. Indiana Univ. Math. J. 21, 1001–1011 (1971/72).

237. J. Fröhlich, C. King, The Chern-Simons Theory and Knot Polynomials. Comm. Math. Phys. 126, 167-199 (1989).

238. J. Fröhlich, E. Seiler, The massive Thirring-Schwinger model (QED2): convergence of perturbation theory and particle structure. Helv. Phys. Acta 49, no. 6, 889–924 (1976).

239. D. Fujiwara, Remarks on convergence of Feynman path integrals, Duke Math. J. 47, 559-600 (1980).

240. D. Fujiwara, The Feynman path integral as an improper integral over the Sobolev space. Journées "Équations aux Dérivées Partielles" (Saint Jean de Monts, 1990), Exp. No. XIV, 15 pp., École Polytech., Palaiseau, (1990).

241. D. Fujiwara, The stationary phase method with an estimate of the remainder term on a space of large dimension. Nagoya Math. J. 124 , 61–97 (1991).

242. D. Fujiwara, Some Feynman path integrals as oscillatory integrals over a Sobolev manifold. Functional analysis and related topics, 1991 (Kyoto), 39–53, Lecture Notes in Math., 1540, Springer, Berlin, (1993).

243. D. Fujiwara, Mathematical Methods for Feynman Path Integrals. (Japanese) Springer, Tokyo, (1999).

244. D. Fujiwara, N. Kumano-go, Smooth functional derivatives in Feynman path integrals by time slicing approximation, Bull. Sci. math. 129, 57–79 (2005).

245. D. Fujiwara, N. Kumano-go, The second term of the semi-classical asymptotic expansion for Feynman path integrals with integrand of polynomial growth. J. Math. Soc. Japan 58, no. 3, 837–867 (2006).

246. D. Fujiwara, N. Kumano-go, Feynman Path Integrals and Semiclassical Approximation. RIMS KoKyuroku Bessatsu B5, 241-263 (2008).

247. D. Fujiwara, T. Tsuchida, The time slicing approximation of the fundamental solution for the Schrödinger equation with electromagnetic fields. J. Math. Soc. Japan 49 (1997), no. 2, 299–327.

248. D. Gatarek, N.Gisin, Continuous quantum jumps and infinite-dimensional stochastic equation, J. Math. Phys. 32, n.8, 2152–2157 (1991).

249. B. Gaveau, L. S. Schulman, Explicit time-dependent Schrödinger propagators. J. Phys. A 19, no. 10, 1833–1846 (1986).

250. T. L. Gill, W. W. Zachary, Foundations for relativistic quantum theory. I. Feynman's operator calculus and the Dyson conjectures. J. Math. Phys. 43 (2002), no. 1, 69–93.

251. R. Gilmore, M. Jeffery, Path integrals and network quantum numbers. On Klauder's path: a field trip, 77–90, World Sci. Publ., River Edge, NJ, (1994).

252. M. J. Goovaerts, A. Babcenco, J. T. Devreese, A new expansion method in the Feynman path integral formalism: Application to a one-dimensional delta-function potential. J. Mathematical Phys. 14, 554–559 (1973).

253. H. Grabert, A. Inomata, L. S. Schulman, U. Weiss, (Editors). Path integrals from meV to MeV: Tutzing '92. Proceedings of the Fourth International Conference held in Tutzing, May 18–21, 1992. World Scientific Publishing Co., Inc., River Edge, NJ, (1993).

254. L. v. Grafenstein, Asymptotische Entwicklungen von Schilder-Integralen und Borel Summation. Dipl. Thesis, University of Bonn (2001).

255. C. Grosche, Path integrals, hyperbolic spaces, and Selberg trace formulae. World Scientific Publishing Co., Inc., River Edge, NJ, (1996).

256. C. Grosche, F. Steiner, Handbook of Feynman path integrals. Springer Tracts in Modern Physics, 145. Springer-Verlag, Berlin, (1998).

257. L. Gross, Abstract Wiener Spaces, Proc. 5th Berkeley Symp. Math. Stat. Prob. 2, 31-42, (1965).

258. M. Grothaus, D.C. Khandekar, J.L. da Silva, L. Streit, The Feynman integral for time-dependent anharmonic oscillators. J. Math. Phys. 38, no. 6, 3278–3299 (1997).

259. M. Grothaus, L. Streit, Quadratic actions, semi-classical approximation, and delta sequences in Gaussian analysis. Rep. Math. Phys. 44, no. 3, 381–405 (1999).

260. M. Grothaus, L. Streit, I.V. Volovich, Knots, Feynman diagrams and Matrix models. Infin. Dimens. Anal. Quantum Probab. Relat. Top. 2, no. 3, 359–380 (1999).

261. M. Grothaus, A. Vogel, The Feynman integrand as a white noise distribution for the Green's function for nonperturbative accessible potentials. in preparation (2008).

262. E. Guadagnini, The link invariants of the Chern-Simons field theory. New developments in topological quantum field theory. de Gruyter Expositions in Mathematics, 10. Walter de Gruyter & Co., Berlin, (1993).

263. M. C. Gutzwiller, Chaos in classical and quantum mechanics. Interdisciplinary Applied Mathematics, 1. Springer-Verlag, New York, (1990).

264. M. C. Gutzwiller, A. Inomata, J. R. Klauder and L. Streit. (Editors). Path integrals from meV to MeV. Papers from the symposium held at the University of Bielefeld, Bielefeld, August 5–9, 1985. Bielefeld Encounters in Physics and Mathematics, VII. World Scientific Publishing Co., Singapore, (1986).

265. Z. Haba, Feynman integral and random dynamics in quantum physics. A probabilistic approach to quantum dynamics. Mathematics and its Applications, 480. Kluwer Academic Publishers, Dordrecht, 1999.

266. A. Hahn, Chern-Simons Theory on $\mathbb{R}^3$ in axial gauge, PhD Thesis, Bonner Schriften Nr. 345 (2001).

267. A. Hahn, Rigorous State Model Representations for the Wilson loop observables in Chern-Simons theory, SFB-611-Preprint, Bonn (2002).

268. A. Hahn, Chern-Simons theory on $\mathbb{R}^3$ in axial gauge: a rigorous approach, J. Funct. Anal. 211 (2), 483–507 (2004).

269. A. Hahn, The Wilson loop observables of Chern-Simons theory on $\mathbb{R}^3$ in axial gauge, Commun. Math. Phys. 248 (3), 467–499 (2004).

270. A. Hahn, Chern-Simons models on $S^2 \times S^1$, torus gauge fixing, and link invariants I, J. Geom. Phys. 53 (3), 275–314 (2005).

271. A. Hahn, An analytic approach to Turaevs shadow invariants, SFB 611 preprint, Bonn, submitted to Commun. Math. Phys.

272. A. Hahn, White noise analysis, quantum field theory, and topology. Stochastic analysis: classical and quantum, 13–30, World Sci. Publ., Hackensack, NJ, (2005).

273. B. Helffer, Semi-classical analysis for the Schrödinger operator and applications. Lecture Notes in Mathematics, 1336. Springer-Verlag, Berlin, (1988).

274. R. Henstock, Integration in product spaces, including Wiener and Feynman integration, Proc. London Math. Soc. 3rd series, 27, 317–344 (1973).

275. T. Hida, Stationary stochastic processes, Mathematical Notes, Princeton University Press, Princeton, N.J.; University of Tokyo Press, Tokyo, (1970).

276. T. Hida, White noise approach to Feynman integrals. J. Korean Math. Soc. 38, no. 2, 275–281 (2001).

277. T. Hida, H.H. Kuo, J. Potthoff, L. Streit, White Noise. An infinite-dimensional calculus. Mathematics and its Applications, 253. Kluwer Academic Publishers Group, Dordrecht, (1993).

278. L. Hörmander, Fourier integral operators I, Acta Math. 127(1), 79–183 (1971).

279. L. Hörmander, The Analysis of Linear Partial Differential Operators, I. Distribution Theory and Fourier Analysis, Springer-Verlag, Berlin/Heidelberg/New York/Tokyo, (1983).

280. Y.Z. Hu, Calculation of Feynman path integral for certain central forces. Stochastic analysis and related topics (Oslo, 1992), 161–171, Stochastics Monogr. 8, Gordon and Breach, Montreux, (1993).

281. Y.Z. Hu, P.A. Meyer, Chaos de Wiener et intégrale de Feynman. (French) [Wiener chaos and the Feynman integral] Séminaire de Probabilités, XXII, 51–71, Lecture Notes in Math., 1321, Springer, Berlin, (1988).

282. N. Hurt, R. Hermann, Quantum statistical mechanics and Lie group harmonic analysis. Part A. Lie Groups: History, Frontiers and Applications, X. Math Sci Press, Brookline, Mass., (1980).

283. T. Ichinose, Path integral for the Dirac equation in two space-time dimensions. Proc. Japan Acad. Ser. A Math. Sci. 58, no. 7, 290–293, (1982).

284. W. Ichinose, A note on the existence and $\hbar$-dependency of the solution of equations in quantum mechanics. Osaka J. Math. 32, no. 2, 327–345 (1995).

285. W. Ichinose, On the formulation of the Feynman path integral through broken line paths. Comm. Math. Phys. 189, no. 1, 17–33 (1997).

286. W. Ichinose, Convergence of the Feynman path integral in the weighted Sobolev spaces and teh representation of correlation functions. J. Math. Soc. Japan 55, no. 4, 957–983 (2003).

287. W. Ichinose, A mathematical theory of the phase space Feynman path integral of the functional. Comm. Math. Phys. 265, no. 3, 739–779 (2006).

288. T. Ichinose, H. Tamura, Propagation of a Dirac particle. A path integral approach. J. Math. Phys. 25, no. 6, 1810–1819 (1984).

289. T. Ichinose, H. Tamura, Sharp error bound on norm convergence of exponential product formula and approximation to kernels of Schrödinger semigroups. Comm. Partial Differential Equations 29, no. 11-12, 1905–1918 (2004).

290. N. Ikeda, S. Manabe, Asymptotic formulae for stochastic oscillatory integrals. Asymptotic problems in probability theory: Wiener functionals and asymptotics (Sanda/Kyoto, 1990), 136–155, Pitman Res. Notes Math. Ser., 284, Longman Sci. Tech., Harlow, (1993).

291. N. Ikeda, I. Shigekawa, S. Taniguchi, The Malliavin calculus and long time asymptotics of certain Wiener integrals. Miniconference on linear analysis and function spaces (Canberra, 1984), 46–113, Proc. Centre Math. Anal. Austral. Nat. Univ., 9, Austral. Nat. Univ., Canberra, (1985).

292. A. Inomata, G. Junker, Quasiclassical path-integral approach to supersymmetric quantum mechanics. Phys. Rev. A (3) 50 , no. 5, 3638–3649 (1994).

293. A. Inomata, G. Junker, Path integrals and Lie groups. Noncompact Lie groups and some of their applications (San Antonio, TX, 1993), 199–224, NATO Adv. Sci. Inst. Ser. C Math. Phys. Sci., 429, Kluwer Acad. Publ., Dordrecht, (1994).

294. A. Inomata, G. Junker, A history of Feynman's sum over histories in quantum mechanics. Path integrals from peV to TeV (Florence, 1998), 19–21, World Sci. Publ., River Edge, NJ, (1999).

295. A. Inoue, A partial solution for Feynman's problem: a new derivation of the Weyl equation. Proceedings of the Symposium on Mathematical Physics and Quantum Field Theory (Berkeley, CA, 1999), 121–145 (electronic), Electron. J. Differ. Equ. Conf., 4, Southwest Texas State Univ., San Marcos, TX, (2000).

296. C. Itzykson, J.B. Zuber, Quantum field theory. International Series in Pure and Applied Physics. McGraw-Hill International Book Co., New York, (1980).

297. G.W. Johnson, The equivalence of two approaches to Feynman integral, J. Math. Phys. 23, 2090–2096, (1982).

298. G.W. Johnson, G. Kallianpur, Homogeneous chaos, p-forms, scaling and the Feynman integral. Trans. Amer. Math. Soc. 340, no. 2, 503–548 (1993).

299. G.W. Johnson, J. G. Kim, A dominated-type convergence theorem for the Feynman integral. J. Math. Phys. 41 , no. 5, 3104–3112 (2000).

300. G.W. Johnson, M.L. Lapidus, Generalized Dyson series, generalized Feynman diagrams, the Feynman integral and Feynman's operational calculus. Mem. Amer. Math. Soc. 62 , no. 351, 1–78 (1986)

301. G.W. Johnson, M.L. Lapidus, Noncommutative operations on Wiener functionals and Feynman's operational calculus. J. Funct. Anal. 81 , no. 1, 74–99 (1988).

302. G.W. Johnson, M.L. Lapidus, The Feynman integral and Feynman's operational calculus, Oxford University Press, New York (2000).

303. G. Jona-Lasinio, Stochastic processes and quantum mechanics. Colloquium in honor of Laurent Schwartz, Vol. 2 (Palaiseau, 1983). Astrisque No. 132 (1985), 203–216.

304. M. Kac, Integration in function spaces and some of its applications. Lezioni Fermiane. [Fermi Lectures] Accademia Nazionale dei Lincei, Pisa, (1980).

305. G. Kallianpur, Traces, natural extensions and Feynman distributions. Gaussian random fields (Nagoya, 1990), 14–27, Ser. Probab. Statist., 1, World Sci. Publ., River Edge, NJ, (1991).

306. G. Kallianpur, C. Bromley, Generalized Feynman integrals using analytic continuation in several complex variables. Stochastic analysis and applications, 217–267, Adv. Probab. Related Topics, 7, Dekker, New York, (1984).

307. G. Kallianpur, D. Kannan, R.L. Karandikar, Analytic and sequential Feynman integrals on abstract Wiener and Hilbert spaces, and a Cameron Martin Formula, Ann. Inst. H. Poincaré, Prob. Th. 21, 323–361 (1985).

308. G. Kallianpur, A. S. Üstünel, Distributions, Feynman integrals and measures on abstract Wiener spaces. Stochastic analysis and related topics (Silivri, 1990), 237–284, Progr. Probab., 31, Birkhuser Boston, Boston, MA, (1992).

309. H. Kanasugi, H. Okada, Systematic Treatment of General Time-Dependent Harmonic Oscillator in Classical and Quantum Mechanics, Progr. Theoret. Phys. 16 (2), 384–388 (1975).

310. L. H. Kauffman, Vassiliev invariants and functional integration without integration. Stochastic analysis and mathematical physics (SAMP/ANESTOC 2002), 91–114, World Sci. Publishing, River Edge, NJ, (2004).

311. L. H. Kauffman, Functional integration and the theory of knots. J. Math. Phys. 36, no. 5, 2402–2429 (1995).

312. L. H. Kauffman, An introduction to knot theory and functional integrals. Functional integration (Cargèse, 1996), 247–308, NATO Adv. Sci. Inst. Ser. B Phys., 361, Plenum, New York, (1997).

313. L. H. Kauffman, Witten's integral and the Kontsevich integral. Particles, fields, and gravitation (Łódź, 1998), 368–381, AIP Conf. Proc., 453, Amer. Inst. Phys., Woodbury, NY, (1998).

314. D. C. Khandekar, S. V. Lawande, Exact propagator for a time-dependent harmonic oscillator with and without a singular perturbation, J. Math. Phys. 16 (2), 384–388 (19975).

315. D. C. Khandekar, S. V. Lawande, K. V. Bhagwat, Path-integral methods and their applications. World Scientific Publishing Co., Inc., River Edge, NJ, (1993).

316. A. Yu. Khrennikov, Feynman measure in phase space and the symbols for infinite-dimensional pseudodiferential operators, Mat. Zam. 37, 734–742 (1986).

317. T. Kinoshita edt.. Quantum Elecrodynamics World scientific, Singapore (1990).

318. J.R. Klauder, Functional techniques and their application in quantum field theory, in "Lectures in Theoretical Physics", Vol. XIVB, W. Britten Ed., Colorado Assoc. Univ. Press (1974).

319. H. Kleinert, Path integrals in quantum mechanics, statistics, polymer physics, and financial markets. Third edition. World Scientific Publishing Co., Inc., River Edge, NJ, (2004).

320. V. N. Klepikov, Asymptotic expansion of the Feynman integral. (Russian) Translated in J. Soviet Math. 53 (1991), no. 1, 29–37. Theory of random processes, No. 14, 29–37, 117, "Naukova Dumka", Kiev, (1986).

321. I. Kluvanek, Integration and the Feynman-Kac formula. Studia Math. 86, no. 1, 35–57 (1987).

322. V. N. Kolokoltsov, Long time behavior of continuously observed and controlled quantum systems (a study of the Belavkin quantum filtering equation). Quantum probability communications, 229–243, QP-PQ, X, World Sci. Publ., River Edge, NJ, (1998).

323. V. N. Kolokoltsov, Complex measures on path space: an introduction to the Feynman integral applied to the Schrödinger equation. Methodol. Comput. Appl. Probab. 1 (1999), no. 3, 349–365.

324. V. N. Kolokoltsov, Semiclassical analysis for diffusions and stochastic processes, Lecture Notes in Mathematics, 1724. Springer-Verlag, Berlin, (2000).

325. V. N. Kolokoltsov, Mathematics of the Feynman path integral. Filomat No. 15, 293–311 (2001).

326. V. N. Kolokoltsov, Small diffusion and fast dying out asymptotics for super-processes as non-Hamiltonian quasiclassics for evolution equations. Electron. J. Probab. 6 , no. 21, 16 pp. (electronic), (2001).

327. V. N. Kolokoltsov, A new path integral representation for the solutions of the Schrödinger, heat and stochastic Schrödinger equations. Math. Proc. Cambridge Philos. Soc. 132, no. 2, 353–375 (2002).

328. V. N. Kolokoltsov, K. Makarov, Asymptotic spectral analysis of a small diffusion operator and the life times of the corresponding diffusion process. Russian J. Math. Phys. 4, no. 3, 341–360 (1996).

329. V. N. Kolokoltsov, A. E. Tyukov, Small time and semiclassical asymptotics for stochastic heat equations driven by Lévy noise. Stoch. Stoch. Rep. 75, no. 1-2, 1–38 (2003).

330. Yu. Kondratiev, P. Leukert, J. Potthoff, L. Streit, W. Westerkamp, Generalized functionals in Gaussian spaces: the characterization theorem revisited, J. Funct. Anal. 141, no. 2, 301–318 (1996).

331. A. L. Koshkarov, The saddle-point method for a functional integral. (Russian) Teoret. Mat. Fiz. 102 , no. 2, 210–216 (1995); translation in Theoret. and Math. Phys. 102, no. 2, 153–157 (1995).

332. A.A. Kostantinov, V.P. Maslov, A.M. Chebotarev, Probability representations of solutions of the Cauchy problem for quantum mechanical equations, Russian Math. Surveys 45 (6), 1–26 (1990).

333. D. V. Ktitarev, R. S. Yegikian, Feynman path integral for Dirac system with analytic potential. J. Math. Phys. 34, no. 7, 2821–2826 (1993).

334. S. Kuksin, The Eulerian limit for 2D statistical hydrodynamics. J. Statist. Phys. 115 , no. 1-2, 469–492 (2004).

335. N. Kumano-go, Feynman path integrals as analysis on path space by time slicing approximation, Bull. Sci. Math. 128 (3), 197–251 (2004).

336. T. Kuna, L. Streit, W. Westerkamp, Feynman integrals for a class of exponentially growing potentials, J. Math. Phys. 39, no. 9, 4476–4491 (1998).

337. H.H. Kuo, Gaussian Measures in Banach Spaces, Lecture Notes in Math., Springer-Verlag Berlin-Heidelberg-New York, (1975).

338. N. P. Landsman, Mathematical topics between classical and quantum mechanics. Springer Monographs in Mathematics. Springer-Verlag, New York, (1998).

339. M.L. Lapidus, The Feynman-Kac formula with a Lebesgue-Stieltjes measure and Feynman's operational calculus. Stud. Appl. Math. 76 , no. 2, 93–132 (1987).

340. M.L. Lapidus, The Feynman integral and Feynman's operational calculus: a heuristic and mathematical introduction. Ann. Math. Blaise Pascal 3, no. 1, 89–102 (1996).

341. M.L. Lapidus, In search of the Riemann zeros, AMS, Providence (2008).

342. M.L. Lapidus, M. van Frankenhuijsen, Fractal geometry and number theory. Complex dimensions of fractal strings and zeros of zeta functions. Birkhuser Boston, Inc., Boston, MA, (2000).

343. M.L. Lapidus, M. van Frankenhuijsen, Fractal geometry, complex dimensions and zeta functions. Geometry and spectra of fractal strings. Springer Monographs in Mathematics. Springer, New York, (2006).

344. B. Lascar, Le noyau de l'équation des ondes sur une variété riemannienne compacte comme intégrale de chemins. (French) J. Analyse Math. 37, 1–31 (1980).

345. B. Lascar, R. Lascar, Équation des ondes pour un opérateur hypo-elliptique à caractéristiques doubles sur une variété symplectique et asymptotiques géométriques. (French) [Wave equation for a hypoelliptic operator with double characteristics on a symplectic manifold and geometric asymptotics] J. Anal. Math. 93, 1–34 (2004).

346. B. Lascar, J. Sjöstrand, Équation de Schrödinger et propagation des singularités pour des opérateurs pseudo-différentiels à caractéristiques réelles de multiplicité variable. (French) [Schrödinger equation and propagation of singularities for pseudodifferential operators with real characteristics of variable multiplicity] Astrisque, 95, 167–207, Astérisque, 95, Soc. Math. France, Paris, (1982).

347. B. Lascar, J. Sjöstrand, Équation de Schrödinger et propagation pour des O.D.P. à caractéristiques relles de multiplicité variable. II. (French) [Schrödinger equation and propagation for pseudodifferential operators with real characteristics of variable multiplicity. II] Comm. Partial Differential Equations 10, no. 5, 467–523 (1985).

348. A. Lascheck, P. Leukert, L. Streit, W. Westerkamp, Quantum Mechanical Propagators in Terms of Hida Distribution, Rep. Math. Phys. 33, 221–232 (1993).

349. R. Léandre, Theory of distribution in the sense of Connes-Hida and Feynman path integral on a manifold. Infin. Dimens. Anal. Quantum Probab. Relat. Top. 6, no. 4, 505–517 (2003).

350. R. Léandre, Deformation quantization in white noise analysis. SIGMA Symmetry Integrability Geom. Methods Appl. 3 (2007), Paper 027, 8 pp. (electronic).

351. R. Léandre, J.R. Norris, Integration by parts and Cameron-Martin formulas for the free path space of a compact Riemannian manifold. Séminaire de Probabilités, XXXI, 16–23, Lecture Notes in Math., 1655, Springer, Berlin, (1997).

352. R. Léandre, A. Rogers, Equivariant cohomology, Fock space and loop groups. J. Phys. A 39 (2006), no. 38, 11929–11946.

353. S. Leukert, J. Schäfer, A Rigorous Construction of Abelian Chern-Simons Path Integral Using White Noise Analysis, Reviews in Math. Phys. 8, 445–456 (1996).

354. W. Lim, V. Sa-yakanit, Bose-Einstein condensation of atomic hydrogen Feynman path integral approach. J. Phys. A 37, no. 19, 5251–5259 (2004).

355. Yu. Yu. Lobanov, O. V. Zeinalova, E. P. Zhidkov, Approximation formulas for functional integrals in $P(\varphi)_2$-quantum field theory. Communications of the Joint Institute for Nuclear Research. Dubna, E11-91-352. Joint Inst. Nuclear Res., Dubna, (1991).

356. K. Loo, A rigorous real time Feynman path integral. J. Math. Phys. 40, no. 1, 64–70 (1999).

357. K. Loo, Rigorous real-time Feynman path integral for vector potentials. J. Phys. A 33, no. 50, 9205–9214 (2000).

358. K. Loo, A rigorous real-time Feynman path integral and propagator. J. Phys. A 33, no. 50, 9215–9239 (2000).

359. P. Malliavin, Sur certaines intégrales stochastiques oscillantes. (French) [Some stochastic oscillatory integrals] C. R. Acad. Sci. Paris Sr. I Math. 295, no. 3, 295–300 (1982).

360. P. Malliavin, S. Taniguchi, Analytic functions, Cauchy formula, and stationary phase on a real abstract Wiener space. J. Funct. Anal. 143, no. 2, 470–528 (1997).

361. V. Mandrekar, Some remarks on various definitions of Feynman integrals, in Lectures Notes Math., Eds K. Jacob Beck (1983), pp.170-177.

362. M. Manoliu, Abelian Chern-Simons theory. II. A functional integral approach. J. Math. Phys. 39, no. 1, 207–217 (1998).

363. V.P. Maslov, Théorie des perturbations et méthodes asymptotiques Dunod, Paris, (1972).

364. V.P. Maslov, Méthodes Opérationelles, Mir. Moscou (1987).

365. V.P. Maslov, Complex Markov chains and a Feynman path integral for nonlinear equations. (Russian). Nauka, Moscow, (1976).

366. V.P. Maslov, Characteristics of pseudo-differential operators, Proc. Internat. Congress Math. (Nice 1970) vol 2, Gautier-Villars, Paris (1971).

367. V.P. Maslov, A.M. Chebotarev, Processus à sauts et leur application dans la mécanique quantique, In: S. Albeverio et al.(ed) Feynman path integrals. Springer Lecture Notes in Physics 106, 58–72 (1979).

368. V.P. Maslov, A.M. Chebotarev, Continual integral over branching trajectories. (Russian) Teoret. Mat. Fiz. 45, no. 3, 329–345 (1980).

369. V.P. Maslov, A.M. Chebotarev, On the second term of the logarithmic asymptotic expansion of functional integrals. (Russian) Probability theory. Mathematical statistics. Theoretical cybernetics, Vol. 19, pp. 127–154, Akad. Nauk SSSR, Vsesoyuz. Inst. Nauchn. i Tekhn. Informatsii, Moscow, (1982).

370. V.P. Maslov, M.V. Fedoryuk, Semiclassical approximation in quantum mechanics. Translated from the Russian by J. Niederle and J. Tolar. Mathematical Physics and Applied Mathematics, 7. Contemporary Mathematics, 5. D. Reidel Publishing Co., Dordrecht-Boston, Mass., (1981).

371. S. Mazzucchi, Feynman Path Integrals, Ph.D Thesis, University of Trento, (2003).

372. M.B. Mensky, Continuous Quantum Measurements and Path Integrals. Taylor & Francis, Bristol and Philadelphia (1993).

373. M.B. Mensky, Quantum measurements and decoherence. Models and phenomenology, Fundamental Theories of Physics, 110. Kluwer Academic Publishers, Dordrecht, (2000).

374. I. Mitoma, One loop approximation of the Chern-Simons integral, Acta Appl. Math. 63, 253–273 (2000).

375. M. M. Mizrahi, The semiclassical expansion of the anharmonic-oscillator propagator. J. Math. Phys. 20, no. 5, 844–855 (1979).

376. M.I. Monastyrskii, Appendix to F.J. Dyson's paper: Missed opportunities, Russ. Math. Surv. 35, n.1, 199–208 (1980).

377. P. Muldowney, A general theory of integration in function spaces, including Wiener and Feynman integration. Pitman Research Notes in Mathematics Series, 153. Longman Scientific & Technical, Harlow; John Wiley & Sons, Inc., New York, (1987).

378. H.J.W. Müller-Kirsten, Introduction to quantum mechanics. Schrödinger equation and path integral. World Scientific Publishing Co. Pte. Ltd., Hackensack, NJ, (2006).

379. T. Nakamura, A nonstandard representation of Feynman's path integrals. J. Math. Phys. 32, no. 2, 457–463 (1991).

380. T. Nakamura, Path space measures for Dirac and Schrödinger equations: nonstandard analytical approach. J. Math. Phys. 38, no. 8, 4052–4072 (1997).

381. T. Nakamura, Path space measure for the $3 + 1$-dimensional Dirac equation in momentum space. J. Math. Phys. 41, no. 8, 5209–5222, (2000).

382. V. E. Nazaikinskii, B.-W. Schulze, B. Yu. Sternin, Quantization methods in differential equations. Differential and Integral Equations and Their Applications, 3. Taylor & Francis, Ltd., London, (2002).

383. E. Nelson, Feynman integrals and the Schrödinger equation, J. Math. Phys. 5, 332–343 (1964).

384. B. Nelson, B. Sheeks, Path integration for velocity-dependent potentials. Comm. Math. Phys. 84, no. 4, 515–530 (1982).

385. F. Nevanlinna, Zur Theorie der asymptotischen Potenzreihen, Ann. Acad. Sci. Fenn. (A) 12 (3), 1–81 (1919).

386. D. Nualart, V. Steblovskaya, Asymptotics of oscillatory integrals with quadratic phase function on Wiener space. Stochastics Stochastics Rep. 66 (1999), no. 3–4, 293–309.

387. F. W. J. Olver, Asymptotics and special functions. Computer Science and Applied Mathematics. Academic Press, New York-London, (1974).

388. G. Papanicolaou, On the convergence of the Feynman path integrals for a certain class of potentials, J. Math. Phys. 31, 342–347 (1990).

389. G. Parisi, Statistical Field Theory, Addison-Wesley, Redwood City, Calif. (1988).

390. P. Pechukas, J. C. Light, On the Exponential Form of Time-Displacement Operators in Quantum Mechanics, J. Chem. Phys. 44 (10), 3897–3912 (1966).

391. V. I. Piterbarg, V. R. Fatalov, The Laplace method for probability measures in Banach spaces. Russian Math. Surveys 50 , no. 6, 1151–1239 (1995)

392. J. Potthoff, L. Streit, A characterization of Hida distributions, J. Funct. Anal. 101, no. 1, 212–229 (1991).

393. P. Protter, Stochastic Integration and Differential Equations, Applications of Mathematics 21, Springer-Verlag (1990).

394. P. Ramond, Field theory: a modern primer. Second edition. Frontiers in Physics, 74. Addison-Wesley Publishing Company, Advanced Book Program, Redwood City, CA, (1990).

395. M. Reed, B. Simon, Methods of Modern Mathematical Physics. Fourier Analysis, Self-Adjointness, Academic Press, New York, (1975).

396. J. Rezende, The method of stationary phase for oscillatory integrals on Hilbert spaces, Comm. Math. Phys. 101, 187–206 (1985).

397. J. Rezende, Quantum systems with time dependent harmonic part and the Morse index, J. Math. Phys. 25 (11), 3264–3269 (1984).

398. J. Rezende, Time-Dependent Linear Hamiltonian Systems and Quantum Mechanics, Lett. Math. Phys. 38, 117–127 (1996).

399. J. Rezende, Feynman integrals and Fredholm determinants, J. Math. Phys. 35 (8), 4357–4371 (1994).

400. D. Robert, Autour de l'approximation semi-classique. (French) [On semiclassical approximation] Progress in Mathematics, 68. Birkhuser Boston, Inc., Boston, MA, (1987).
401. G. Roepstorff, Path integral approach to quantum physics. An introduction. Texts and Monographs in Physics. Springer-Verlag, Berlin, (1994).
402. A. Rogers, Supersymmetric path integration. Phys. Lett. B 193 (1987), no. 1, 48–54.
403. A. Rogers, Supermanifolds. Theory and applications. World Scientific Publishing Co. Pte. Ltd., Hackensack, NJ, (2007).
404. V. Sa-yakanit, W.Sritrakool, J.-O. Berananda et al. (Editors). Path integrals from meV to MeV. Proceedings of the Third International Conference held in Bangkok, January 9–13, 1989. World Scientific Publishing Co., Inc., Teaneck, NJ, (1989).
405. J. Schäfer, Abelsche Chern-Simons Theorie - eine Mathematische Konstruktion, Diplomarbeit Bochum (1991).
406. L. S. Schulman, Techniques and applications of path integration. A Wiley-Interscience Publication. John Wiley & Sons, Inc., New York, (1981).
407. A. S. Schwartz, The partition function of a degenerate quadratic functional and Ray-Singer invariants. Lett. Math. Phys 2, 247–252 (1978).
408. A. S. Schwartz, The partition function of a degenerate functional. Comm. Math. Phys. 67, no. 1, 1–16 (1979).
409. G. L. Sewell, Quantum mechanics and its emergent macrophysics. Princeton University Press, Princeton, NJ, (2002).
410. N. N. Shamarov, A functional integral with respect to a countably additive measure representing a solution of the Dirac equation. Trans. Moscow Math. Soc., 243–255 (2005).
411. E. T. Shavgulidze, Solutions of Schrödinger equations with polynomial potential by path integrals. Path integrals: Dubna '96, 269–272, Joint Inst. Nuclear Res., Dubna, 1996.
412. B. Simon, Trace ideals and their applications, London Mathematical Society Lecture Note Series 35. Cambridge U.P., Cambridge, (1979).
413. B. Simon, Functional integration and quantum physics, Second edition. AMS Chelsea Publishing, Providence, RI, (2005).
414. J. L. Silva, L. Streit, Feynman integrals and white noise analysis. Stochastic analysis and mathematical physics (SAMP/ANESTOC 2002), 285–303, World Sci. Publ., River Edge, NJ, (2004).
415. M. Sirugue, M. Sirugue-Collin, A. Truman, Semiclassical approximation and microcanonical ensemble. Ann. Inst. H. Poincar Phys. Thor. 41 , no. 4, 429–444 (1984).
416. W. Słowikowski, The use of fenoms in classical analysis on Hilbert spaces, Aarhus University Preprint Series, Mathem. Inst. 1974/75, No 26.
417. A. Smailagic, E. Spallucci, Feynman path integral on the non-commutative plane. J. Phys. A 36, no. 33, L467–L471 (2003).
418. O.G. Smolyanov, A.Yu. Khrennikov, The central limit theorem for generalized measures of infinite dimensional spaces, Sov. Math. Dokl. 31, 301-304 (1985).
419. O.G. Smolyanov, E.T. Shavgulidze, Representation of the solutions of second-order linear evolution superdifferential equations by path integrals. (Russian) Dokl. Akad. Nauk SSSR 309, no. 3, 545–550 (1989).
420. O.G. Smolyanov, E.T. Shavgulidze, Path integrals, (Russian). Moskov. Gos. Univ., Moscow, (1990).

421. O.G. Smolyanov, E.T. Shavgulidze, Feynman formulas for solutions of infinite-dimensional Schrödinger equations with polynomial potentials. (Russian) Dokl. Akad. Nauk 390, no. 3, 321–324 (2003).

422. O.G. Smolyanov, M.O. Smolyanova, Transformations of the Feynman integral under nonlinear transformations of the phase space. (Russian) Teoret. Mat. Fiz. 100 (1994), no. 1, 3–13; translation in Theoret. and Math. Phys. 100 (1994), no. 1, 803–810 (1995).

423. A.D. Sokal, An improvement of Watson's theorem on Borel summability, J. Math. Phys. 21, no. 2, 261–263 (1980).

424. E. M. Stein, Harmonic analysis: real-variable methods, orthogonality, and oscillatory integrals. Princeton Mathematical Series, 43. Monographs in Harmonic Analysis, III. Princeton University Press, Princeton, NJ, (1993).

425. R. F. Streater, Euclidean quantum mechanics and stochastic integrals. Stochastic integrals (Proc. Sympos., Univ. Durham, Durham, 1980), pp. 371–393, Lecture Notes in Math., 851, Springer, Berlin-New York, (1981).

426. L. Streit, Feynman paths, sticky walls, white noise. A garden of quanta, 105–113, World Sci. Publ., River Edge, NJ, (2003).

427. L. Streit, T. Hida, Generalized Brownian functional and the Feynman integral. Stoch. Proc. Appl. 16, 55–69 (1984).

428. H. Sugita, S. Taniguchi, Oscillatory integrals with quadratic phase function on a real abstract Wiener space. J. Funct. Anal. 155, no. 1, 229–262 (1998).

429. H. Sugita, S. Taniguchi, A remark on stochastic oscillatory integrals with respect to a pinned Wiener measure. Kyushu J. Math. 53, no. 1, 151–162 (1999).

430. T. Sunada, Homology, Wiener integrals, and integrated densities of states. J. Funct. Anal. 106, no. 1, 50–58 (1992).

431. M. Suzuki, General theory of fractal path integrals with applications to many-body theories and statistical physics. J. Math. Phys. 32, no. 2, 400–407 (1991).

432. F. Takeo, Generalized vector measures and path integrals for hyperbolic systems. Hokkaido Math. J. 18, no. 3, 497–511 (1989).

433. S. Taniguchi, On the exponential decay of oscillatory integrals on an abstract Wiener space. J. Funct. Anal. 154, no. 2, 424–443 (1998).

434. S. Taniguchi, A remark on stochastic oscillatory integrals with respect to a pinned Wiener measure. Kyushu J. Math. 53, no. 1, 151–162 (1999).

435. S. Taniguchi, Lévy's stochastic area and the principle of stationary phase. J. Funct. Anal. 172, no. 1, 165–176 (2000).

436. S. Taniguchi, Exponential decay of stochastic oscillatory integrals on classical Wiener spaces. J. Math. Soc. Japan 55, no. 1, 59–79 (2003).

437. S. Taniguchi, Stochastic oscillatory integrals: asymptotics and exact expressions for quadratic phase function. Stochastic analysis and mathematical physics (SAMP/ANESTOC 2002), 165–181, World Sci. Publ., River Edge, NJ, (2004).

438. S. Taniguchi, On the Jacobi field approach to stochastic oscillatory integrals with quadratic phase function. Kyushu J. Math. 61, no. 1, 191–208 (2007).

439. T. J. S. Taylor, Applications of harmonic analysis on the infinite-dimensional torus to the theory of the Feynman integral. Functional integration with emphasis on the Feynman integral (Sherbrooke, PQ, 1986). Rend. Circ. Mat. Palermo (2) Suppl. No. 17 (1987), 349–362 (1988).

440. H. Thaler, Solution of Schrödinger equations on compact Lie groups via probabilistic methods, Potential Anal. 18 no. 2, 119–140 (2003).

441. H. Thaler, The Doss Trick on Symmetric Spaces, Letters in Mathematical Physics 72, 115–127 (2005).
442. E. G. Thomas, Path integrals on finite sets. Acta Appl. Math. 43, no. 2, 191–232 (1996).
443. E. G. Thomas, Finite path integrals. Proceedings of the Norbert Wiener Centenary Congress, 1994 (East Lansing, MI, 1994), 225–232, Proc. Sympos. Appl. Math., 52, Amer. Math. Soc., Providence, RI, (1997).
444. E. G. Thomas, Path distributions on sequence spaces. Infinite dimensional stochastic analysis (Amsterdam, 1999), 235–268, Verh. Afd. Natuurkd. 1. Reeks. K. Ned. Akad. Wet., 52, R. Neth. Acad. Arts Sci., Amsterdam, (2000).
445. E. G. Thomas, Path integrals associated with Sturm-Liouville operators. J. Korean Math. Soc. 38, no. 2, 365–383 (2001).
446. E. G. Thomas, Projective limits of complex measures and martingale convergence. Probab. Theory Related Fields 119, no. 4, 579–588 (2001).
447. A. Truman, Classical mechanics, the diffusion (heat) equation, and the Schrödinger equation. J. Math. Phys. 18, no. 12, 2308–2315 (1977).
448. A. Truman, The Feynman maps and the Wiener integral, J. Math. Phys. 19, 1742–1750 (1978).
449. A. Truman, The polygonal path formulation of the Feynman path integral, In: S. Albeverio et al. (ed) Feynman path integrals. Springer Lecture Notes in Physics 106, (1979).
450. A. Truman, T. Zastawniak, Stochastic PDE's of Schrödinger type and stochastic Mehler kernels—a path integral approach, Seminar on Stochastic Analysis, Random Fields and Applications (Ascona, 1996), 275–282, Birkhäuser, Basel, (1999).
451. A. Truman, T. Zastawniak, Stochastic Mehler kernels via oscillatory path integrals, J. Korean Math. Soc. 38(2), 469–483 (2001).
452. A. Truman, H. Z. Zhao, The stochastic Hamilton Jacobi equation, stochastic heat equations and Schrödinger equations. Stochastic analysis and applications (Powys, 1995), 441–464, World Sci. Publ., River Edge, NJ, (1996).
453. A. Truman, H. Z. Zhao, Stochastic Burgers' equations and their semi-classical expansions. Comm. Math. Phys. 194, no. 1, 231–248 (1998).
454. A. Truman, H. Z. Zhao, Semi-classical limit of wave functions. Proc. Amer. Math. Soc. 128 , no. 4, 1003–1009 (2000).
455. A. Truman, H. Z. Zhao, Burgers equation and the WKB-Langer asymptotic L^2 approximation of eigenfunctions and their derivatives. Probabilistic methods in fluids, 332–366, World Sci. Publ., River Edge, NJ, (2003).
456. T. Tsuchida, Remarks on Fujiwara's stationary phase method on a space of large dimension with a phase function involving electromagnetic fields. Nagoya Math. J. 136, 157–189 (1994).
457. A. N. Varchenko, Newton polyhedra and estimates of oscillatory integrals. (Russian) Funkcional. Anal. Appl. 10, no. 3, 13–38 (1976).
458. A. N. Vasiliev, Functional methods in quantum field theory and statistical physics. Gordon and Breach Science Publishers, Amsterdam, (1998).
459. H. Watanabe, Path integral for some systems of partial differential equations. Proc. Japan Acad. Ser. A Math. Sci. 60, no. 3, 86–89 (1984).
460. H. Watanabe, Feynman-Kac formula associated with a system of partial differential operators. J. Funct. Anal. 65, no. 2, 204–235 (1986).

461. K. D. Watling, Formulae for solutions to (possibly degenerate) diffusion equations exhibiting semi-classical asymptotics. Stochastics and quantum mechanics (Swansea, 1990), 248–271, World Sci. Publ., River Edge, NJ, (1992).

462. S. Weinberg, The quantum theory of fields. Vol. I. Modern applications. Cambridge University Press, Cambridge, (2005).

463. S. Weinberg, The quantum theory of fields. Vol. II. Foundations. Cambridge University Press, Cambridge, (2005).

464. F. W. Wiegel, Introduction to path-integral methods in physics and polymer science. World Scientific Publishing Co., Singapore, (1986).

465. E. Witten, Quantum field theory and the Jones polynomial, Commun. Math. Phys. 121, 353–389 (1989).

466. K. Yajima, Smoothness and Non-Smoothness of the Fundamental Solution of Time Dependent Schrödiger Equations, Commun. Math. Phys. 181, 605–629 (1996).

467. J.C. Zambrini, Feynman integrals, diffusion processes and quantum symplectic two-forms. J. Korean Math. Soc. 38 , no. 2, 385–408 (2001).

468. T. J. Zastawniak, Approximation of Feynman path integrals by integrals over finite dimensional spaces, Bull. Pol. Acad. Sci. Math. 34, 355–372 (1986).

469. T. J. Zastawniak, Path integrals for the telegrapher's and Dirac equations; the analytic family of measures and the underlying Poisson process. Bull. Polish Acad. Sci. Math. 36 (1988), no. 5-6, 341–356 (1989).

470. T. J. Zastawniak, Path integrals for the Dirac equation—some recent developments in mathematical theory. Stochastic analysis, path integration and dynamics (Warwick, 1987), 243–263, Pitman Res. Notes Math. Ser., 200, Longman Sci. Tech., Harlow, (1989).

471. T. J. Zastawniak, The nonexistence of the path-space measure for the Dirac equation in four space-time dimensions. J. Math. Phys. 30, no. 6, 1354–1358 (1989).

472. T. J. Zastawniak, Equivalence of Albeverio. Høegh-Krohn-Feynman integral for anharmonic oscillators and the analytic Feynman integral, Univ. Iagel. Acta Math. 28, 187–199 (1991).

473. T. J. Zastawniak, Fresnel type path integral for the stochastic Schrödinger equation. Lett. Math. Phys. 41, no. 1, 93–99 (1997).

474. R. L. Zimmerman, Evaluation of Feynman's functional integrals. J. Mathematical Phys. 6, 1117–1124 (1965).

475. J. Zinn-Justin, Quantum field theory and critical phenomena. Second edition. International Series of Monographs on Physics, 85. Oxford Science Publications. The Clarendon Press, Oxford University Press, New York, (1993).

Analytic Index

Abstract Wiener spaces, 104, 106

Action, Hamilton's principle of least
    action, 2, 6, 108

Analytic continuation
    of Wiener integrals, 108

Analytic continuation, definition of
    Feynman integrals by, 4, 62,
    115–116
    of Wiener integrals, 4, 115

Anharmonic oscillator, 7, 37, 51, 58, 93
    Dynamics by Fresnel integrals, 61
    Expectations with respect to the
      Gibbs state of the harmonic
      oscillator, 69
    Expectations with respect to the
      ground state of the harmonic
      oscillator, 63
    Feynman–Ito formula, small times,
      38
    Fresnel integrals, small times, 38
    Solutions of Schrödinger equation
      by Fresnel integrals, 57, 58

Annihilation-creation operators, 73, 74

Automorphisms of Weyl algebra, 71, 75,
    76, 78, 82, 87, 88, 91

Banach function algebra, over $\mathbb{R}^n$, 10
    over $\mathcal{H}$, 12
    over D^*, 40

Borel summable asymptotic expansions,
    110

Boson fields, 6, 85–91

Brownian motion, 4, 5

Causal propagator, 88

Chern–Simons functional integral, 120,
    136–138

Classical action, 6, 19, 63, 73, 79, 85,
    88, 108, 113, 131–133

Classical limit, 4, 5, 108–115

Commutation relations, 71, 74, 75

Complex measure, 4

Composition with entire functions, 11,
    12

Cyclic vector, 76

Dirac equation, 140

Euclidean–Markov fields, 6

Euler–Lagrange equations, 1

Expectations with respect to the
      ground state of the harmonic
      oscillator, 63
    for relativistic quantum boson
      fields, 85–91
    with respect to invariant quasi free
      states, 83
    with respect to the Gibbs state of
      the harmonic oscillator, 69–71

Exponential interactions, 7, 90–91

Feynman history integrals, heuristic,
    6–8
    as Fresnel integrals relative to a
      quadratic form, 85–91

Feynman path integral
    as infinite dimensional oscillatory
      integrals:

anharmonic oscillator, 98
quartic oscillator, 105–107
definition by "analytic continua-
    tion", 62
other definitions, 115, 120–124
Feynman path integral, definition by
    "analytic continuation", 4,
    115–116
  as Fresnel integrals:
    anharmonic oscillators, 51–71
    anharmonic oscillators, small
        times, 37, 38
    infinitely many harmonic
        oscillators, 73–83
    non relativistic quantum
        mechanics, 19–35
    relativistic quantum fields,
        85–91
    scattering operator, 32–35
    solutions of Schrödinger
        equation, 19–24
    wave operators, 24–32
  definition by "sequential limit", 4
  Other definitions, 15–16
Feynman path integral, definition by
    Poisson processes, 123–124
Feynman path integral, definition by
    sequential approach, 120–123
Feynman path integral, definition by
    white noise, 116–120
Feynman path integral,definition by
    sequential approach, 123
Feynman–Ito formula, 19, 23
Feynman–Kac formula, 19, 115
Fields, *see* Boson fields
Finitely based function, 12
Fock representation, 74, 83, 88
Fractals, 140
Free Boson field, 85
Fresnel integrable, on $\mathbb{R}^n$, 10, 11
  on a real Hilbert space, 12
Fresnel integrals, on $\mathbb{R}^n$, 9–11
  Computations of, 19–24, 27–32, 54,
    67–71, 81–83, 88–91
  for anharmonic oscillators, small
    times, 37–38
  applications, 51–91
  projectics, 40–50

relative to a non degenerate
    quadratic form, definition,
    38–40
in non relativistic quantum me-
    chanics (potential scattering),
    19–35
on a real separable Banach space,
    47–50
on a real separable Hilbert space,
    definition, 11–12
on a real separable Hilbert space,
    properties, 12–16
Fubini theorem, for Fresnel integrals on
    real Hilbert space, 14–15
  for Fresnel integrals on real
    separable Banach space, 49
  for Fresnel integrals relative to a
    quadratic form, 42–50

Gaussian measure, 16
Gentle perturbations, 28
Gibbs states, 7, 83, 88
Green's function, 21, 53, 59, 126, 128,
    129

Hamilton's principle, 2, 6, 108
Harmonic oscillator, finitely many
    degrees of freedom, 63–73
  infinitely many degrees of freedom,
    73–83, 85–91
  time dependent frequency, 127–129
Heat equation, 3
Homogeneous boundary condition,
    transformation to, 52
Hyperbolic systems, 140

Improper normalized integral, 28, 32
Indefinite metric, 16
Infinite dimensional oscillatory integrals
    class p normalized, 97
  normalized with respect to an
    operator B, 96
Initial value problem, 19
Integrals, *see* Feynman Integrals,
    Fresnel Integrals, etc.
Integration in function spaces, 6
Invariance, Fresnel integral, under
    nearly isometric transforma-
    tions, 16

Invariance, orthogonal, for normalized
        integral, 12–13
Invariance, translation, for normalized
        integral, 9, 12–13
    for normalized integral with
        respect to quadratic form, 40,
        57
Invariant quasi free states, 75
Ito's functional, 16

Laplace method, 116

Magnetic fields, 124–130
Maslov index, 113
Models quantum fields, 6–7, 86–91

Non commutative, 140
Non relativistic quantum mechanics,
        1–5, 19–35, 51–83, 105, 108
Non standard analysis, 138
Normalization, 9
Normalization integral, see Fresnel
        integral

Oscillatory integrals, see Fresnel
        integrals, Feynman history,
        Feynman path integrals,
        93–100
    on $\mathbb{R}^n$, 94
    on real separable Hilbert spaces,
        95
    with polynomial phase function,
        101–105

Parseval relation (Path integrals, see
        Feynman-, Wiener-), 11–16
Parseval type formula, 95, 97, 98, 103,
        105, 134
Periodic solutions, 114
Phase space Feynman path integrals,
        130–132, 140
Poisson processes, 123–124
Polynomial interactions, 7, 105
Potential scattering, 19–35

Quantum mechanics, non relativistic,
        19–35, 51–83, 105, 108
    Relativistic, 85–91

Statistical, 69–71, 140
Quantum theory of measurement, 133
Quasi-free states, 73, 75, 83, 87, 91

Scattering amplitude, scattering
        operator, 32
Schrödinger equation, 105
Schrödinger equation, semiclassical
        limit, 112
Schrödinger equation, solutions given
        by Fresnel integrals, 19–24,
        51–58
Schrödinger equation, solutions given
        by infinite dimensional
        oscillatory integrals, 98–100,
        106
    Magnetic field, 124
    manifolds, 116, 140
    time dependent potentials, 125–130
Semiclassical expansion, 108–115, 139
Semigroup, 61
Sequential approach, 120–123
Series expansions, 20–35, 53–62, 67–71,
        91, 108–115
Sine–Gordon interactions, 90
State on the Weyl algebra, 71–90
Stationary phase, 108–115
Stochastic field theory, 7
Stochastic mechanics, 5
Stochastic Schrödinger equation,
        133–135
Supersymmetric, 140
Symplectic space, 76, 78
Symplectic transformation, 76, 78, 87

Time dependent potentials, 120, 125
Time ordered vacuum expectation
        values, 6
Trace of the Schrödinger group, 114

Wave operator, 24–34
Weyl algebra, 71, 74, 82, 87, 88, 90
White noise calculus, 116–120
White noise distributions, 118
White noise, definition of Feynman
        integrals by, 116–120
Wiener's measure, 4, 5

List of Notations

Symbols

$\hbar$ Planck's constant divided by 2π, 2

$\int_{\mathcal{H}}^{\triangle} e^{\frac{i}{2}(x,Bx)} f(x)\mathrm{d}(x)$ Fresnel integral with respect to quadratic form $\triangle$, or integral normalized with respect to $\triangle$, 39

$\widetilde{\int} e^{\frac{i}{2}|x|^2} f(x)\mathrm{d}x$, $\widetilde{\int_{\mathcal{H}}} e^{\frac{i}{2}|x|^2} f(x)\mathrm{d}x$ normalized integral or Fresnel integral (on space $\mathcal{H}$), 10, 11

$\widetilde{\int_{\mathcal{H}}} e^{\frac{i}{2}\langle x,x\rangle} f(x)\mathrm{d}x$ Fresnel or normalized integral on Banach space $(E, <, >)$, 47

$\widetilde{\int_{\gamma(t)=x}}$ "Feynman path integral" notation for certain Fresnel integrals over spaces of paths, 22, 23, 37, 51, 55, 58

$\widetilde{\int} \lim_{t\to\pm\infty} \frac{\gamma(t)}{t}=V_{\pm}$ notation for certain (improper) Fresnel integrals of scattering theory, 33, 34

$\int_{\mathbb{R}^n}^{\circ} e^{i\Phi(x)} f(x)\mathrm{d}x$ oscillatory integral of f with respect to the phase function Φ, 94

$\sigma \vee \tau = \max\{\sigma, \tau\}$, 21

$\widetilde{\int_{\mathbb{R}^n}^{\circ}} e^{\frac{i}{2\hbar}|x|^2} f(x)\mathrm{d}x$ oscillatory integral of f with quadratic phase function, 94

$\widetilde{\int_{\mathcal{H}}^{p}} e^{\frac{i}{2\hbar}|x|^2} e^{-\frac{i}{2\hbar}(x,Lx)} f(x)\mathrm{d}x$ class p normalized oscillatory integral of f with respect to the operator L, 98

$\widetilde{\int_{\mathcal{H}}^{\circ}} e^{\frac{i}{2\hbar}|x|^2} f(x)\mathrm{d}x$ infinite dimensional oscillatory integral of f on the Hilbert space $\mathcal{H}$, 95

$\widetilde{\int_{\mathcal{H}}^{B}} e^{\frac{i}{2\hbar}(x,Bx)} f(x)\mathrm{d}x$ normalized oscillatory integral of f with respect to the operator B, 96

$s \wedge t = \min\{s,t\}$, 25

Letters

$C(\mathcal{H})$ (continuous bounded functions on $\mathcal{H}$), 40

$\mathcal{M}(\mathbb{R}^n), \mathcal{M}(\mathcal{H})$ spaces of complex measures, 9, 11

$\mathcal{S}(\mathbb{R}^n)$ Schwartz space, 2

$\Phi(\vec{x}), \Phi(f)$ quantized time zero free field, 87

$\mathbb{R}^n$ Euclidean n dimensional space, 1

$\mathcal{F}(\mathbb{R}^n), \mathcal{F}(\mathcal{H}), \mathcal{F}(D^*)$ space of Fresnel integrable functions, 11, 39

$\mathcal{F}(f)$ Fresnel integral, 10, 12

$\mathcal{F}_{\triangle}(f)$ Fresnel integral relative to quadratic form $\triangle$, 40

$\mathcal{H}$ real separable Hilbert space, 11

m particle mass, 1

Lecture Notes in Mathematics

For information about earlier volumes
please contact your bookseller or Springer
LNM Online archive: springerlink.com

Vol. 1755: J. Azéma, M. Émery, M. Ledoux, M. Yor (Eds.), Séminaire de Probabilités XXXV (2001)

Vol. 1756: P. E. Zhidkov, Korteweg de Vries and Nonlinear Schrödinger Equations: Qualitative Theory (2001)

Vol. 1757: R. R. Phelps, Lectures on Choquet's Theorem (2001)

Vol. 1758: N. Monod, Continuous Bounded Cohomology of Locally Compact Groups (2001)

Vol. 1759: Y. Abe, K. Kopfermann, Toroidal Groups (2001)

Vol. 1760: D. Filipović, Consistency Problems for Heath-Jarrow-Morton Interest Rate Models (2001)

Vol. 1761: C. Adelmann, The Decomposition of Primes in Torsion Point Fields (2001)

Vol. 1762: S. Cerrai, Second Order PDE's in Finite and Infinite Dimension (2001)

Vol. 1763: J.-L. Loday, A. Frabetti, F. Chapoton, F. Goichot, Dialgebras and Related Operads (2001)

Vol. 1764: A. Cannas da Silva, Lectures on Symplectic Geometry (2001)

Vol. 1765: T. Kerler, V. V. Lyubashenko, Non-Semisimple Topological Quantum Field Theories for 3-Manifolds with Corners (2001)

Vol. 1766: H. Hennion, L. Hervé, Limit Theorems for Markov Chains and Stochastic Properties of Dynamical Systems by Quasi-Compactness (2001)

Vol. 1767: J. Xiao, Holomorphic Q Classes (2001)

Vol. 1768: M. J. Pflaum, Analytic and Geometric Study of Stratified Spaces (2001)

Vol. 1769: M. Alberich-Carramiñana, Geometry of the Plane Cremona Maps (2002)

Vol. 1770: H. Gluesing-Luerssen, Linear Delay-Differential Systems with Commensurate Delays: An Algebraic Approach (2002)

Vol. 1771: M. Émery, M. Yor (Eds.), Séminaire de Probabilités 1967-1980. A Selection in Martingale Theory (2002)

Vol. 1772: F. Burstall, D. Ferus, K. Leschke, F. Pedit, U. Pinkall, Conformal Geometry of Surfaces in S^4 (2002)

Vol. 1773: Z. Arad, M. Muzychuk, Standard Integral Table Algebras Generated by a Non-real Element of Small Degree (2002)

Vol. 1774: V. Runde, Lectures on Amenability (2002)

Vol. 1775: W. H. Meeks, A. Ros, H. Rosenberg, The Global Theory of Minimal Surfaces in Flat Spaces. Martina Franca 1999. Editor: G. P. Pirola (2002)

Vol. 1776: K. Behrend, C. Gomez, V. Tarasov, G. Tian, Quantum Comohology. Cetraro 1997. Editors: P. de Bartolomeis, B. Dubrovin, C. Reina (2002)

Vol. 1777: E. García-Río, D. N. Kupeli, R. Vázquez-Lorenzo, Osserman Manifolds in Semi-Riemannian Geometry (2002)

Vol. 1778: H. Kiechle, Theory of K-Loops (2002)

Vol. 1779: I. Chueshov, Monotone Random Systems (2002)

Vol. 1780: J. H. Bruinier, Borcherds Products on $O(2,1)$ and Chern Classes of Heegner Divisors (2002)

Vol. 1781: E. Bolthausen, E. Perkins, A. van der Vaart, Lectures on Probability Theory and Statistics. Ecole d' Eté de Probabilités de Saint-Flour XXIX-1999. Editor: P. Bernard (2002)

Vol. 1782: C.-H. Chu, A. T.-M. Lau, Harmonic Functions on Groups and Fourier Algebras (2002)

Vol. 1783: L. Grüne, Asymptotic Behavior of Dynamical and Control Systems under Perturbation and Discretization (2002)

Vol. 1784: L. H. Eliasson, S. B. Kuksin, S. Marmi, J.-C. Yoccoz, Dynamical Systems and Small Divisors. Cetraro, Italy 1998. Editors: S. Marmi, J.-C. Yoccoz (2002)

Vol. 1785: J. Arias de Reyna, Pointwise Convergence of Fourier Series (2002)

Vol. 1786: S. D. Cutkosky, Monomialization of Morphisms from 3-Folds to Surfaces (2002)

Vol. 1787: S. Caenepeel, G. Militaru, S. Zhu, Frobenius and Separable Functors for Generalized Module Categories and Nonlinear Equations (2002)

Vol. 1788: A. Vasil'ev, Moduli of Families of Curves for Conformal and Quasiconformal Mappings (2002)

Vol. 1789: Y. Sommerhäuser, Yetter-Drinfel'd Hopf algebras over groups of prime order (2002)

Vol. 1790: X. Zhan, Matrix Inequalities (2002)

Vol. 1791: M. Knebusch, D. Zhang, Manis Valuations and Prüfer Extensions I: A new Chapter in Commutative Algebra (2002)

Vol. 1792: D. D. Ang, R. Gorenflo, V. K. Le, D. D. Trong, Moment Theory and Some Inverse Problems in Potential Theory and Heat Conduction (2002)

Vol. 1793: J. Cortés Monforte, Geometric, Control and Numerical Aspects of Nonholonomic Systems (2002)

Vol. 1794: N. Pytheas Fogg, Substitution in Dynamics, Arithmetics and Combinatorics. Editors: V. Berthé, S. Ferenczi, C. Mauduit, A. Siegel (2002)

Vol. 1795: H. Li, Filtered-Graded Transfer in Using Noncommutative Gröbner Bases (2002)

Vol. 1796: J.M. Melenk, hp-Finite Element Methods for Singular Perturbations (2002)

Vol. 1797: B. Schmidt, Characters and Cyclotomic Fields in Finite Geometry (2002)

Vol. 1798: W.M. Oliva, Geometric Mechanics (2002)

Vol. 1799: H. Pajot, Analytic Capacity, Rectifiability, Menger Curvature and the Cauchy Integral (2002)

Vol. 1800: O. Gabber, L. Ramero, Almost Ring Theory (2003)

Vol. 1801: J. Azéma, M. Émery, M. Ledoux, M. Yor (Eds.), Séminaire de Probabilités XXXVI (2003)

Vol. 1802: V. Capasso, E. Merzbach, B. G. Ivanoff, M. Dozzi, R. Dalang, T. Mountford, Topics in Spatial Stochastic Processes. Martina Franca, Italy 2001. Editor: E. Merzbach (2003)

Vol. 1803: G. Dolzmann, Variational Methods for Crystalline Microstructure – Analysis and Computation (2003)
Vol. 1804: I. Cherednik, Ya. Markov, R. Howe, G. Lusztig, Iwahori-Hecke Algebras and their Representation Theory. Martina Franca, Italy 1999. Editors: V. Baldoni, D. Barbasch (2003)
Vol. 1805: F. Cao, Geometric Curve Evolution and Image Processing (2003)
Vol. 1806: H. Broer, I. Hoveijn. G. Lunther, G. Vegter, Bifurcations in Hamiltonian Systems. Computing Singularities by Gröbner Bases (2003)
Vol. 1807: V. D. Milman, G. Schechtman (Eds.), Geometric Aspects of Functional Analysis. Israel Seminar 2000-2002 (2003)
Vol. 1808: W. Schindler, Measures with Symmetry Properties (2003)
Vol. 1809: O. Steinbach, Stability Estimates for Hybrid Coupled Domain Decomposition Methods (2003)
Vol. 1810: J. Wengenroth, Derived Functors in Functional Analysis (2003)
Vol. 1811: J. Stevens, Deformations of Singularities (2003)
Vol. 1812: L. Ambrosio, K. Deckelnick, G. Dziuk, M. Mimura, V. A. Solonnikov, H. M. Soner, Mathematical Aspects of Evolving Interfaces. Madeira, Funchal, Portugal 2000. Editors: P. Colli, J. F. Rodrigues (2003)
Vol. 1813: L. Ambrosio, L. A. Caffarelli, Y. Brenier, G. Buttazzo, C. Villani, Optimal Transportation and its Applications. Martina Franca, Italy 2001. Editors: L. A. Caffarelli, S. Salsa (2003)
Vol. 1814: P. Bank, F. Baudoin, H. Föllmer, L.C.G. Rogers, M. Soner, N. Touzi, Paris-Princeton Lectures on Mathematical Finance 2002 (2003)
Vol. 1815: A. M. Vershik (Ed.), Asymptotic Combinatorics with Applications to Mathematical Physics. St. Petersburg, Russia 2001 (2003)
Vol. 1816: S. Albeverio, W. Schachermayer, M. Talagrand, Lectures on Probability Theory and Statistics. Ecole d'Eté de Probabilités de Saint-Flour XXX-2000. Editor: P. Bernard (2003)
Vol. 1817: E. Koelink, W. Van Assche (Eds.), Orthogonal Polynomials and Special Functions. Leuven 2002 (2003)
Vol. 1818: M. Bildhauer, Convex Variational Problems with Linear, nearly Linear and/or Anisotropic Growth Conditions (2003)
Vol. 1819: D. Masser, Yu. V. Nesterenko, H. P. Schlickewei, W. M. Schmidt, M. Waldschmidt, Diophantine Approximation. Cetraro, Italy 2000. Editors: F. Amoroso, U. Zannier (2003)
Vol. 1820: F. Hiai, H. Kosaki, Means of Hilbert Space Operators (2003)
Vol. 1821: S. Teufel, Adiabatic Perturbation Theory in Quantum Dynamics (2003)
Vol. 1822: S.-N. Chow, R. Conti, R. Johnson, J. Mallet-Paret, R. Nussbaum, Dynamical Systems. Cetraro, Italy 2000. Editors: J. W. Macki, P. Zecca (2003)
Vol. 1823: A. M. Anile, W. Allegretto, C. Ringhofer, Mathematical Problems in Semiconductor Physics. Cetraro, Italy 1998. Editor: A. M. Anile (2003)
Vol. 1824: J. A. Navarro González, J. B. Sancho de Salas, $\mathscr{C}^{\infty}$ – Differentiable Spaces (2003)
Vol. 1825: J. H. Bramble, A. Cohen, W. Dahmen, Multiscale Problems and Methods in Numerical Simulations, Martina Franca, Italy 2001. Editor: C. Canuto (2003)
Vol. 1826: K. Dohmen, Improved Bonferroni Inequalities via Abstract Tubes. Inequalities and Identities of Inclusion-Exclusion Type. VIII, 113 p, 2003.

Vol. 1827: K. M. Pilgrim, Combinations of Complex Dynamical Systems. IX, 118 p, 2003.
Vol. 1828: D. J. Green, Gröbner Bases and the Computation of Group Cohomology. XII, 138 p, 2003.
Vol. 1829: E. Altman, B. Gaujal, A. Hordijk, Discrete-Event Control of Stochastic Networks: Multimodularity and Regularity. XIV, 313 p, 2003.
Vol. 1830: M. I. Gil', Operator Functions and Localization of Spectra. XIV, 256 p, 2003.
Vol. 1831: A. Connes, J. Cuntz, E. Guentner, N. Higson, J. E. Kaminker, Noncommutative Geometry, Martina Franca, Italy 2002. Editors: S. Doplicher, L. Longo (2004)
Vol. 1832: J. Azéma, M. Émery, M. Ledoux, M. Yor (Eds.), Séminaire de Probabilités XXXVII (2003)
Vol. 1833: D.-Q. Jiang, M. Qian, M.-P. Qian, Mathematical Theory of Nonequilibrium Steady States. On the Frontier of Probability and Dynamical Systems. IX, 280 p, 2004.
Vol. 1834: Yo. Yomdin, G. Comte, Tame Geometry with Application in Smooth Analysis. VIII, 186 p, 2004.
Vol. 1835: O.T. Izhboldin, B. Kahn, N.A. Karpenko, A. Vishik, Geometric Methods in the Algebraic Theory of Quadratic Forms. Summer School, Lens, 2000. Editor: J.-P. Tignol (2004)
Vol. 1836: C. Năstăsescu, F. Van Oystaeyen, Methods of Graded Rings. XIII, 304 p, 2004.
Vol. 1837: S. Tavaré, O. Zeitouni, Lectures on Probability Theory and Statistics. Ecole d'Eté de Probabilités de Saint-Flour XXXI-2001. Editor: J. Picard (2004)
Vol. 1838: A.J. Ganesh, N.W. O'Connell, D.J. Wischik, Big Queues. XII, 254 p, 2004.
Vol. 1839: R. Gohm, Noncommutative Stationary Processes. VIII, 170 p, 2004.
Vol. 1840: B. Tsirelson, W. Werner, Lectures on Probability Theory and Statistics. Ecole d'Eté de Probabilités de Saint-Flour XXXII-2002. Editor: J. Picard (2004)
Vol. 1841: W. Reichel, Uniqueness Theorems for Variational Problems by the Method of Transformation Groups (2004)
Vol. 1842: T. Johnsen, A. L. Knutsen, K₃ Projective Models in Scrolls (2004)
Vol. 1843: B. Jefferies, Spectral Properties of Noncommuting Operators (2004)
Vol. 1844: K.F. Siburg, The Principle of Least Action in Geometry and Dynamics (2004)
Vol. 1845: Min Ho Lee, Mixed Automorphic Forms, Torus Bundles, and Jacobi Forms (2004)
Vol. 1846: H. Ammari, H. Kang, Reconstruction of Small Inhomogeneities from Boundary Measurements (2004)
Vol. 1847: T.R. Bielecki, T. Björk, M. Jeanblanc, M. Rutkowski, J.A. Scheinkman, W. Xiong, Paris-Princeton Lectures on Mathematical Finance 2003 (2004)
Vol. 1848: M. Abate, J. E. Fornaess, X. Huang, J. P. Rosay, A. Tumanov, Real Methods in Complex and CR Geometry, Martina Franca, Italy 2002. Editors: D. Zaitsev, G. Zampieri (2004)
Vol. 1849: Martin L. Brown, Heegner Modules and Elliptic Curves (2004)
Vol. 1850: V. D. Milman, G. Schechtman (Eds.), Geometric Aspects of Functional Analysis. Israel Seminar 2002-2003 (2004)
Vol. 1851: O. Catoni, Statistical Learning Theory and Stochastic Optimization (2004)
Vol. 1852: A.S. Kechris, B.D. Miller, Topics in Orbit Equivalence (2004)
Vol. 1853: Ch. Favre, M. Jonsson, The Valuative Tree (2004)

Vol. 1854: O. Saeki, Topology of Singular Fibers of Differential Maps (2004)

Vol. 1855: G. Da Prato, P.C. Kunstmann, I. Lasiecka, A. Lunardi, R. Schnaubelt, L. Weis, Functional Analytic Methods for Evolution Equations. Editors: M. Iannelli, R. Nagel, S. Piazzera (2004)

Vol. 1856: K. Back, T.R. Bielecki, C. Hipp, S. Peng, W. Schachermayer, Stochastic Methods in Finance, Bressanone/Brixen, Italy, 2003. Editors: M. Fritelli, W. Runggaldier (2004)

Vol. 1857: M. Émery, M. Ledoux, M. Yor (Eds.), Séminaire de Probabilités XXXVIII (2005)

Vol. 1858: A.S. Cherny, H.-J. Engelbert, Singular Stochastic Differential Equations (2005)

Vol. 1859: E. Letellier, Fourier Transforms of Invariant Functions on Finite Reductive Lie Algebras (2005)

Vol. 1860: A. Borisyuk, G.B. Ermentrout, A. Friedman, D. Terman, Tutorials in Mathematical Biosciences I. Mathematical Neurosciences (2005)

Vol. 1861: G. Benettin, J. Henrard, S. Kuksin, Hamiltonian Dynamics – Theory and Applications, Cetraro, Italy, 1999. Editor: A. Giorgilli (2005)

Vol. 1862: B. Helffer, F. Nier, Hypoelliptic Estimates and Spectral Theory for Fokker-Planck Operators and Witten Laplacians (2005)

Vol. 1863: H. Führ, Abstract Harmonic Analysis of Continuous Wavelet Transforms (2005)

Vol. 1864: K. Efstathiou, Metamorphoses of Hamiltonian Systems with Symmetries (2005)

Vol. 1865: D. Applebaum, B.V. R. Bhat, J. Kustermans, J. M. Lindsay, Quantum Independent Increment Processes I. From Classical Probability to Quantum Stochastic Calculus. Editors: M. Schürmann, U. Franz (2005)

Vol. 1866: O.E. Barndorff-Nielsen, U. Franz, R. Gohm, B. Kümmerer, S. Thorbjønsen, Quantum Independent Increment Processes II. Structure of Quantum Lévy Processes, Classical Probability, and Physics. Editors: M. Schürmann, U. Franz, (2005)

Vol. 1867: J. Sneyd (Ed.), Tutorials in Mathematical Biosciences II. Mathematical Modeling of Calcium Dynamics and Signal Transduction. (2005)

Vol. 1868: J. Jorgenson, S. Lang, $Pos_n(R)$ and Eisenstein Series. (2005)

Vol. 1869: A. Dembo, T. Funaki, Lectures on Probability Theory and Statistics. Ecole d'Eté de Probabilités de Saint-Flour XXXIII-2003. Editor: J. Picard (2005)

Vol. 1870: V.I. Gurariy, W. Lusky, Geometry of Müntz Spaces and Related Questions. (2005)

Vol. 1871: P. Constantin, G. Gallavotti, A.V. Kazhikhov, Y. Meyer, S. Ukai, Mathematical Foundation of Turbulent Viscous Flows, Martina Franca, Italy, 2003. Editors: M. Cannone, T. Miyakawa (2006)

Vol. 1872: A. Friedman (Ed.), Tutorials in Mathematical Biosciences III. Cell Cycle, Proliferation, and Cancer (2006)

Vol. 1873: R. Mansuy, M. Yor, Random Times and Enlargements of Filtrations in a Brownian Setting (2006)

Vol. 1874: M. Yor, M. Émery (Eds.), In Memoriam Paul-André Meyer - Séminaire de Probabilités XXXIX (2006)

Vol. 1875: J. Pitman, Combinatorial Stochastic Processes. Ecole d'Eté de Probabilités de Saint-Flour XXXII-2002. Editor: J. Picard (2006)

Vol. 1876: H. Herrlich, Axiom of Choice (2006)

Vol. 1877: J. Steuding, Value Distributions of L-Functions (2007)

Vol. 1878: R. Cerf, The Wulff Crystal in Ising and Percolation Models, Ecole d'Eté de Probabilités de Saint-Flour XXXIV-2004. Editor: Jean Picard (2006)

Vol. 1879: G. Slade, The Lace Expansion and its Applications, Ecole d'Eté de Probabilités de Saint-Flour XXXIV-2004. Editor: Jean Picard (2006)

Vol. 1880: S. Attal, A. Joye, C.-A. Pillet, Open Quantum Systems I, The Hamiltonian Approach (2006)

Vol. 1881: S. Attal, A. Joye, C.-A. Pillet, Open Quantum Systems II, The Markovian Approach (2006)

Vol. 1882: S. Attal, A. Joye, C.-A. Pillet, Open Quantum Systems III, Recent Developments (2006)

Vol. 1883: W. Van Assche, F. Marcellàn (Eds.), Orthogonal Polynomials and Special Functions, Computation and Application (2006)

Vol. 1884: N. Hayashi, E.I. Kaikina, P.I. Naumkin, I.A. Shishmarev, Asymptotics for Dissipative Nonlinear Equations (2006)

Vol. 1885: A. Telcs, The Art of Random Walks (2006)

Vol. 1886: S. Takamura, Splitting Deformations of Degenerations of Complex Curves (2006)

Vol. 1887: K. Habermann, L. Habermann, Introduction to Symplectic Dirac Operators (2006)

Vol. 1888: J. van der Hoeven, Transseries and Real Differential Algebra (2006)

Vol. 1889: G. Osipenko, Dynamical Systems, Graphs, and Algorithms (2006)

Vol. 1890: M. Bunge, J. Funk, Singular Coverings of Toposes (2006)

Vol. 1891: J.B. Friedlander, D.R. Heath-Brown, H. Iwaniec, J. Kaczorowski, Analytic Number Theory, Cetraro, Italy, 2002. Editors: A. Perelli, C. Viola (2006)

Vol. 1892: A. Baddeley, I. Bárány, R. Schneider, W. Weil, Stochastic Geometry, Martina Franca, Italy, 2004. Editor: W. Weil (2007)

Vol. 1893: H. Hanßmann, Local and Semi-Local Bifurcations in Hamiltonian Dynamical Systems, Results and Examples (2007)

Vol. 1894: C.W. Groetsch, Stable Approximate Evaluation of Unbounded Operators (2007)

Vol. 1895: L. Molnár, Selected Preserver Problems on Algebraic Structures of Linear Operators and on Function Spaces (2007)

Vol. 1896: P. Massart, Concentration Inequalities and Model Selection, Ecole d'Été de Probabilités de Saint-Flour XXXIII-2003. Editor: J. Picard (2007)

Vol. 1897: R. Doney, Fluctuation Theory for Lévy Processes, Ecole d'Été de Probabilités de Saint-Flour XXXV-2005. Editor: J. Picard (2007)

Vol. 1898: H.R. Beyer, Beyond Partial Differential Equations, On linear and Quasi-Linear Abstract Hyperbolic Evolution Equations (2007)

Vol. 1899: Séminaire de Probabilités XL. Editors: C. Donati-Martin, M. Émery, A. Rouault, C. Stricker (2007)

Vol. 1900: E. Bolthausen, A. Bovier (Eds.), Spin Glasses (2007)

Vol. 1901: O. Wittenberg, Intersections de deux quadriques et pinceaux de courbes de genre 1, Intersections of Two Quadrics and Pencils of Curves of Genus 1 (2007)

Vol. 1902: A. Isaev, Lectures on the Automorphism Groups of Kobayashi-Hyperbolic Manifolds (2007)

Vol. 1903: G. Kresin, V. Maz'ya, Sharp Real-Part Theorems (2007)

Vol. 1904: P. Giesl, Construction of Global Lyapunov Functions Using Radial Basis Functions (2007)

Vol. 1905: C. Prévôt, M. Röckner, A Concise Course on Stochastic Partial Differential Equations (2007)

Vol. 1906: T. Schuster, The Method of Approximate Inverse: Theory and Applications (2007)

Vol. 1907: M. Rasmussen, Attractivity and Bifurcation for Nonautonomous Dynamical Systems (2007)

Vol. 1908: T.J. Lyons, M. Caruana, T. Lévy, Differential Equations Driven by Rough Paths, Ecole d'Été de Probabilités de Saint-Flour XXXIV-2004 (2007)

Vol. 1909: H. Akiyoshi, M. Sakuma, M. Wada, Y. Yamashita, Punctured Torus Groups and 2-Bridge Knot Groups (I) (2007)

Vol. 1910: V.D. Milman, G. Schechtman (Eds.), Geometric Aspects of Functional Analysis. Israel Seminar 2004-2005 (2007)

Vol. 1911: A. Bressan, D. Serre, M. Williams, K. Zumbrun, Hyperbolic Systems of Balance Laws. Cetraro, Italy 2003. Editor: P. Marcati (2007)

Vol. 1912: V. Berinde, Iterative Approximation of Fixed Points (2007)

Vol. 1913: J.E. Marsden, G. Misiołek, J.-P. Ortega, M. Perlmutter, T.S. Ratiu, Hamiltonian Reduction by Stages (2007)

Vol. 1914: G. Kutyniok, Affine Density in Wavelet Analysis (2007)

Vol. 1915: T. Bıyıkoğlu, J. Leydold, P.F. Stadler, Laplacian Eigenvectors of Graphs. Perron-Frobenius and Faber-Krahn Type Theorems (2007)

Vol. 1916: C. Villani, F. Rezakhanlou, Entropy Methods for the Boltzmann Equation. Editors: F. Golse, S. Olla (2008)

Vol. 1917: I. Veselić, Existence and Regularity Properties of the Integrated Density of States of Random Schrödinger (2008)

Vol. 1918: B. Roberts, R. Schmidt, Local Newforms for GSp(4) (2007)

Vol. 1919: R.A. Carmona, I. Ekeland, A. Kohatsu-Higa, J.-M. Lasry, P.-L. Lions, H. Pham, E. Taflin, Paris-Princeton Lectures on Mathematical Finance 2004. Editors: R.A. Carmona, E. Çinlar, I. Ekeland, E. Jouini, J.A. Scheinkman, N. Touzi (2007)

Vol. 1920: S.N. Evans, Probability and Real Trees. Ecole d'Été de Probabilités de Saint-Flour XXXV-2005 (2008)

Vol. 1921: J.P. Tian, Evolution Algebras and their Applications (2008)

Vol. 1922: A. Friedman (Ed.), Tutorials in Mathematical BioSciences IV. Evolution and Ecology (2008)

Vol. 1923: J.P.N. Bishwal, Parameter Estimation in Stochastic Differential Equations (2008)

Vol. 1924: M. Wilson, Littlewood-Paley Theory and Exponential-Square Integrability (2008)

Vol. 1925: M. du Sautoy, L. Woodward, Zeta Functions of Groups and Rings (2008)

Vol. 1926: L. Barreira, V. Claudia, Stability of Nonautonomous Differential Equations (2008)

Vol. 1927: L. Ambrosio, L. Caffarelli, M.G. Crandall, L.C. Evans, N. Fusco, Calculus of Variations and Non-Linear Partial Differential Equations. Cetraro, Italy 2005. Editors: B. Dacorogna, P. Marcellini (2008)

Vol. 1928: J. Jonsson, Simplicial Complexes of Graphs (2008)

Vol. 1929: Y. Mishura, Stochastic Calculus for Fractional Brownian Motion and Related Processes (2008)

Vol. 1930: J.M. Urbano, The Method of Intrinsic Scaling. A Systematic Approach to Regularity for Degenerate and Singular PDEs (2008)

Vol. 1931: M. Cowling, E. Frenkel, M. Kashiwara, A. Valette, D.A. Vogan, Jr., N.R. Wallach, Representation Theory and Complex Analysis. Venice, Italy 2004. Editors: E.C. Tarabusi, A. D'Agnolo, M. Picardello (2008)

Vol. 1932: A.A. Agrachev, A.S. Morse, E.D. Sontag, H.J. Sussmann, V.I. Utkin, Nonlinear and Optimal Control Theory. Cetraro, Italy 2004. Editors: P. Nistri, G. Stefani (2008)

Vol. 1933: M. Petkovic, Point Estimation of Root Finding Methods (2008)

Vol. 1934: C. Donati-Martin, M. Émery, A. Rouault, C. Stricker (Eds.), Séminaire de Probabilités XLI (2008)

Vol. 1935: A. Unterberger, Alternative Pseudodifferential Analysis (2008)

Vol. 1936: P. Magal, S. Ruan (Eds.), Structured Population Models in Biology and Epidemiology (2008)

Vol. 1937: G. Capriz, P. Giovine, P.M. Mariano (Eds.), Mathematical Models of Granular Matter (2008)

Vol. 1938: D. Auroux, F. Catanese, M. Manetti, P. Seidel, B. Siebert, I. Smith, G. Tian, Symplectic 4-Manifolds and Algebraic Surfaces. Cetraro, Italy 2003. Editors: F. Catanese, G. Tian (2008)

Vol. 1939: D. Boffi, F. Brezzi, L. Demkowicz, R.G. Durán, R.S. Falk, M. Fortin, Mixed Finite Elements, Compatibility Conditions, and Applications. Cetraro, Italy 2006. Editors: D. Boffi, L. Gastaldi (2008)

Vol. 1940: J. Banasiak, V. Capasso, M.A.J. Chaplain, M. Lachowicz, J. Miękisz, Multiscale Problems in the Life Sciences. From Microscopic to Macroscopic. Będlewo, Poland 2006. Editors: V. Capasso, M. Lachowicz (2008)

Vol. 1941: S.M.J. Haran, Arithmetical Investigations. Representation Theory, Orthogonal Polynomials, and Quantum Interpolations (2008)

Vol. 1942: S. Albeverio, F. Flandoli, Y.G. Sinai, SPDE in Hydrodynamic. Recent Progress and Prospects. Cetraro, Italy 2005. Editors: G. Da Prato, M. Röckner (2008)

Vol. 1943: L.L. Bonilla (Ed.), Inverse Problems and Imaging. Martina Franca, Italy 2002 (2008)

Vol. 1944: A. Di Bartolo, G. Falcone, P. Plaumann, K. Strambach, Algebraic Groups and Lie Groups with Few Factors (2008)

Vol. 1945: F. Brauer, P. van den Driessche, J. Wu (Eds.), Mathematical Epidemiology (2008)

Recent Reprints and New Editions

Vol. 1702: J. Ma, J. Yong, Forward-Backward Stochastic Differential Equations and their Applications. 1999 – Corr. 3rd printing (2007)

Vol. 830: J.A. Green, Polynomial Representations of GL_n, with an Appendix on Schensted Correspondence and Littelmann Paths by K. Erdmann, J.A. Green and M. Schoker 1980 – 2nd corr. and augmented edition (2007)

Vol. 1693: S. Simons, From Hahn-Banach to Monotonicity (Minimax and Monotonicity 1998) – 2nd exp. edition (2008)

Vol. 470: R.E. Bowen, Equilibrium States and the Ergodic Theory of Anosov Diffeomorphisms. With a preface by D. Ruelle. Edited by J.-R. Chazottes. 1975 – 2nd rev. edition (2008)

Vol. 523: S.A. Albeverio, R.J. Høegh-Krohn, S. Mazzucchi, Mathematical Theory of Feynman Path Integral. 1976 – 2nd corr. and enlarged edition (2008)